THE LAST TORTOISE

A Tale of Extinction in Our Lifetime

CRAIG B. STANFORD

THE BELKNAP PRESS OF HARVARD UNIVERSITY PRESS

Cambridge, Massachusetts · London, England

2010

Library of Congress Cataloging-in-Publication Data
Stanford, Craig B. (Craig Britton)
The last tortoise : a tale of extinction in our lifetime / Craig B. Stanford.
p.   cm.
Includes bibliographical references and index.
ISBN 978-0-674-04992-5 (alk. paper)
1. Turtles.   2. Endangered species.   I. Title.
QL666.C5S68   2010
333.95′79246—dc22   2009053340

For my family, again

CONTENTS

The Tortoise and the Hare?   1

1   What Exactly Are Tortoises and Turtles?   10

2   Live Long and Prosper   35

3   No Respect for the Ancient Lands   62

4   Eating Tortoises   90

5   "Such Huge Defourmed Creatures"   108

6   Beloved Captives   139

7   Are There Solutions?   163

The Achilles and the Tortoise   181

APPENDIXES

Extremes of the Tortoise World   189
Tortoises on the Brink of Extinction   190
Tortoise Species   191

FURTHER READING   195

ACKNOWLEDGMENTS   200

INDEX   203

The Last Tortoise

THE TORTOISE AND THE HARE?

Tortoises and turtles are at the forefront of the global battle to prevent an imminent mass extinction. Every year more than 10 million tortoises and turtles change hands in Asia; tens of thousands per day. Most of these are taken from the wild or smuggled openly across the borders that divide the nations of Southeast Asia from China. They may have been smuggled first from Indonesia or Malaysia into these border countries. The border, so ironclad when it comes to military issues, is just a porous membrane when illegal trade in wild animals or their parts is involved. The trade in live turtles bound for the Chinese food markets is worth an estimated 700 million dollars annually. There are several varieties of turtles in China that have never been seen alive in the wild; their existence is known only from a few of the last remaining individuals that were spotted in markets. A conservationist who visited one such market in the late 1990s estimated at least ten thousand turtles, most of them taken from the wild, for sale in that market alone.

Freshwater turtles and tortoises are peas in a classification pod; a tortoise is a club-footed, dome-shelled, dry land version of its turtle cousin; the latter uses webbed feet to paddle through freshwater ponds, rivers, and swamps. Both groups are considered equally delicious and so both are critically threatened, although many turtles seem to have a better ability to rebound from human slaughter than most tortoise species. Species whose flesh is not regarded as delicious may be ground into traditional medicines. Although eating turtles and tortoises is an ancient practice in many East Asian countries, in recent years the combination of explosive population growth and newfound consumer cash has created a force so environmentally destructive that the term "crisis," applied liberally at times, fits. Wholesale food companies in China have standing orders in Indonesia, Malaysia, and other countries for thousands of pounds of turtle and tortoise meat per week.

What's more, those species that are not persecuted for the dinner table or the pharmacy shelves are taken by the thousands for the global pet trade. Some of the trade is perfectly legal, if inadvisable, the result of the developing world's eagerness to turn a profit on its natural resources. But much of the trade is black, illegal. A villager in Madagascar for whom a dollar is a day's salary can take a radiated tortoise, resplendent in its starburst shell, and sell it at a profit of five dollars to a local middleman. That tortoise will be smuggled in a suitcase or false-bottomed crate to a distribution point in Asia, perhaps Singapore or Bangkok, and then onward, to be bought by a collector in Europe, North America, or Japan for five thousand dollars. And the illegal trade deficit with Asia has a flipside. An estimated 32 million turtles left the United States in a recent four-year period. Most of these were farm-raised here, not poached, and most were bound for East Asia. But some—the exact percentage is dicey but perhaps three or four percent—were poached from American woodlands and lakes to be smuggled into the Asian food markets. The point is, turtles and tortoises have become a global trade item. In the case of the rarest and most beautiful species—and therefore the most valuable—the black market profits to be made may in some cases rival those from the global trade in narcotics.

There are countries in Southeast Asia that have for millennia been centers of biodiversity which are now emptied of their entire tortoise and turtle fauna. Vietnam is almost devoid of its many species of native tortoises and freshwater turtles—they've all gone to markets in China. Laos, Cambodia, Myanmar, and Indonesia are close behind. Conservationists refer to China as the "black hole" of wildlife conservation for its ongoing appetite for the animals of the entire Asian continent. A nation and an industry hungry to supply an evermore-affluent society of tortoise eaters and pet keepers will not be denied. It is not only Asia that has become a black hole for turtles and tortoises. Africa and the islands of the Indian Ocean are the global center for species diversity of living tortoises, and also the continent on which a battle must be waged to prevent an even greater mass extinction of these small creatures than is currently being carried out in Asia.

Although many of the world's three hundred or so species of tortoises and freshwater and marine turtles are under severe pressure in the early twenty-first century, I focus on one group—the tortoises—in this book. The plight of sea turtles has gotten a great deal of attention in recent years, and although their human-created threats are severe, some populations have shown signs of recovery through years of enlightened management policies coupled with increased public awareness. They have been written about in a plethora of books thanks to the late great biologist Archie Carr and his disciples. But endangered tortoises have had far less such literary good fortune, and their time is about to run out. This book is about the imminent extinction of animals that have walked the Earth for many millions of years. We have long heard the conservationist's cant, "Our grandchildren won't be able to see them alive." In the case of tortoises, we must reassess. Sometime in the next few years they will be gone. The forecast is as dire as it is accurate. Between the time that I write these words and this book reaches bookstores, a few of the species I write about may have been pushed beyond the brink into extinction, the point at which densities in the wild are too low to allow their recovery and the species is doomed without urgent measures.

The real shame is not that it is happening in our own lifetimes, but that it should happen at all. Extinctions are almost always preventable at relatively small human cost. Even near-extinction is reversible—species on the cusp of extermination can often be brought back to healthy population levels. But when the last breeding member of a species dies, that species is dead forever, like some beloved gemstone lost at the bottom of the sea.

Most of my professional life has been devoted to the study of large mammals. Many of them are in danger of extinction too. But I have always been utterly fascinated by and enamored of tortoises. Their tranquil nature; their imperturbable, zen approach to life; their wizened faces; their ancientness give them an almost supernatural air. Tortoises seem controlled by a different force of gravity than the rest of us, with an internal clock that runs on some other-worldly schedule.

No other creature that lives down in the detritus of the Earth's floor has the dignity of an old tortoise. The honorable traits that they possess are carried with aplomb through thick and thin for decades that turn into centuries.

Tortoises are not as majestic as the great whales, as sleekly gorgeous as a tiger, as endearing as a giant panda, or as eerily evocative of ourselves as a mountain gorilla. A tortoise is the ultimate humble, unassuming creature, expecting nothing more than to be left alone deep inside that marvelous shell to live its long life peacefully. They are, nonetheless, perhaps the most astoundingly unlikely creatures in the history of life on planet Earth. If a biologist from another galaxy were to visit our planet on some intergalactic research trip, he would write home about tortoises with surprise and awe. As the famed turtle biologist Archie Carr put it, they have managed the astounding evolutionary feat of relocating their shoulders and hips inside their rib cages and placing a huge protective shield upon their back. Following this radical achievement of anatomical shape-shifting, they have stuck with the program for more than 200 million years. There is no other backboned animal that combines such a unique set of adaptations with a stay-the-course approach to evolutionary success.

There are about forty species of tortoises in the world, plus many racial variants that take the number to around eighty living varieties. A handful more are very recently extinct. They cover a dazzling spectrum of form and function. They range in size from the tiny speckled padloper that barely reaches four inches and a few ounces, to the great tortoises of the oceanic islands of the Pacific and Indian Oceans, which may top five hundred pounds in the wild and break the scales at nearly half a ton in the overfed, under-exercised confines of a zoo. Today these tortoises are linked by their vulnerability to many of the same intransigent threats.

During many years spent studying wild animals, I've seen firsthand the desperate problems they face at the hands of people. We cut down their forests, hunt them for food and profit, take them from their mothers and lock them in shocking conditions of confinement, and generally regard them as food, toys, or threats. In short, they are our expendable commodities, nothing more than pieces of meat or currency exchange. Because studying tropical animals means spending a lot of time in tropical forests, I've become acquainted with many endangered species and the conservation issues confronting them. The case of tortoises and turtles is more critical and poignant than any. Having spent much of my career working with primates, I had thought that their survival prospects in the twenty-first century were grim. That was until I began to see the magnitude and speed with which our shelled kin are being driven to extinction. Their time on Earth, 220 million years and counting, may be over in a blink. There are entire regions of the globe from which these small creatures have been eliminated, vacuumed from the land by the hand of man. There is no group of vertebrates for which extinction risk in the twenty-first century is more real and more imminent.

In this book I detail the risks facing these creatures. I set the scene by describing what a tortoise is and why it is unique in ways that go well beyond that wondrous shell. I explain the role of tortoises in modern ecosystems and the way in which extreme longevity has been their greatest evolutionary breakthrough, though perhaps also their downfall. I build what will seem a bleak picture, describing the many threats to their continued existence. But I also offer in the last chapters some reasons for hope in the face of this seemingly inevitable tragedy. Like many intractable environmental problems, extinction is not manifest destiny. The ozone hole was reduced and global warming can likewise be mitigated. Species loss can be minimized through sensible policies, diligent work to implement the policies, and community-based programs to make local people part of the solution rather than the problem.

Of course many thousands of other creatures are threatened by the hand of humans. But tortoises, with their slow pace of life, their old-man expressions, extreme longevity, and fortress shells speak to us in a way that many other creatures do not. The animal that has come closest to overcoming earthly mortality is now disappearing before our eyes. Every day more habitat is lost. Every day thousands upon thousands are carted from their native lands by people and taken, often clandestinely, to market.

Some of the risk factors have nothing to do with the tortoise's biology. If you burn down a forest to plant a farm, you kill everything in sight except for those very few species that don't need their forest home to survive. But there are many intrinsic factors that predispose animal species to extinction. Some are generalists and can cope, even thrive, in habitats from forest to fields. Biologists call these weed species, for their hard-core adaptability. North American gray squirrels, introduced to Europe, became a weed species, a scourge that has wiped out the native red squirrels in many areas. The red-eared slider, that time-honored pet-store turtle, is now a pest species in many parts of the world. Other species are habitat specialists, supremely adapted to a particular, narrow set of conditions, and utterly at sea when those conditions are removed or altered.

Tortoises have among the most extraordinary life cycles of any animal on Earth. Few if any backboned animals live longer; upward of a century and a half. They may remain as physically spry and reproductively able at a hundred as at twenty, a feat that scientists are struggling to understand. Although the giant tortoise species typically live the longest, even small turtles and tortoises routinely live more than a half-century.

This life-in-the-slow-lane approach is one of the factors that has made existence precarious for tortoises and their kin. A slow reproductive strategy may have advantages in the wild under the right circumstances, but when people are collecting you, eating you, and cutting down your habitat, it's an albatross around the neck. Tortoises are extraordinarily vulnerable for this reason. They are among the first species to disappear from an ecosystem under hunting pressure, being easily found and carried. The hunting pressure faced by tortoises happens for a good reason: they taste good. Although almost any living thing in a tropical forest is edible, tortoises are particularly prized for their flesh, and deep cultural values stand in the way of effective conservation plans that seek to curtail what we have come to call "the bushmeat trade."

Despite these risk factors, most tortoises have withstood human hunting for millennia. Only now have events caught up with them. In setting out to document the threat of extinction facing all wild things, it's useful to use this grand lineage to illustrate how badly things can go wrong at the hand of man. It is also important to see how seemingly insoluble problems can in fact have solutions. The gloom-and-doom attitude many conservationists take toward the state of the natural world is counterproductive. In *The Last Tortoise* I try to show that there are solutions to all but the most dire problems, and very often what seems like a species lost can be turned into a success story with hard work by diligent people unwilling to accept defeat.

Aesop's fable *The Tortoise and the Hare* is a familiar story to us all. In a race between the fleet rabbit and the slow-but-steady tortoise, the tortoise wins by dint of sheer plodding. In reality, the tortoises are losing a race for survival exactly because of their slow, steady pace of life and pace of reproduction. This book explores whether this must inevitably be so. For if it is not written in stone—and it is not—then they can and must be saved.

1

WHAT EXACTLY ARE TORTOISES
AND TURTLES?

The spring bubbles out of the hillside and becomes a stream, flowing downslope for several hundred yards into an egg-shaped lake. I follow the stream on a warm day in late May, flipping and replacing rocks and logs to see the world underfoot. The bank is lined with moss, and dead leaves from the previous autumn form a rug of rotting organic stuff on the forest floor. Through many years working in forests in far-flung corners of the world, I've returned to this place many times: a woodlot in suburban New Jersey. As a teenager I came here to learn about the natural world, to watch the stars, and occasionally to search for a place to be alone with a girlfriend. The remnant of a former expanse of maple and oak forest, today it lives on as a triangle of forest and ponds wedged between freeways and subdivisions. On weekends the sounds of hikers and picnickers are drowned out by the engines of bikers roaring through.

Over the years I've seen the decline of many animal species that were common in the 1970s, even as some other species have skyrocketed. Deer and squirrels, always common, now overrun the place. But some of the salamanders and turtles I caught as an eager-eyed kid are rare to vanishing these days, and the number and diversity of migratory birds in the spring has noticeably declined.

In a glade near the confluence of the stream and its lake, a white-tailed deer doe and fawn are browsing. I once nearly stepped on a newborn fawn in such a place. True to the intended purpose of its spots, the fawn was invisible to me as I picked my way across a rocky slope in dappled sunlight. My foot nearly landed on the fawn. I knelt down slowly and watched her, eyes as big as half-dollars staring at me, body frozen. I squatted by her for long minutes, and only when I stood up did the fawn bolt, bleating distress calls to her mom, a doe who appeared at the edge of the clearing to escort her away.

I'm standing quietly in dappled shade admiring the doe when a noise underfoot pulls my eyes to the ground. A box turtle is ambling along the leafy carpet. He is no more than five inches from nose to tail. Even from six feet above him I can see sharp crimson eyes—a male. Females have fawn-colored eyes. I squat down to have a closer look, and he promptly sucks in his head and legs and slams his shell shut. His carapace—the upper shell—is a domed Jackson Pollock; a splash of orange streaks and blotches against a warm brown background. I turn him over. His plastron—the under shell— is yellow-brown with dark seams and a pair of hinges. Hence the name box turtle, bestowed on the species by early North American naturalists. This is the animal I most sought as a child in these woods. To find one was to connect with the place and with something quasi-magical.

Turtles and Tortoises

There is something supernatural about a turtle. Somewhere deep within this beautifully painted skeleton lie the brain and soul of a timeless creature. Box turtles spend their long, slow lives on the same small patch of ground where they were hatched, and so represent a living link to the past in the same way that ancient trees do. Box turtles have been found rambling in fields with dates carved on them. The carved letters are often worn by time and distorted by the growth of the shell, but sometimes (if we can trust the honesty of the graffiti artist) date the turtle to eighty or a hundred years old. In the mid-1800s, a box turtle was found in a meadow in Connecticut, a few hundred yards from the spot where it had been captured and inscribed by someone sixty years earlier. Two researchers in the 1960s found one box turtle twenty-eight times within twenty-five feet of its original capture location many years earlier. The discovery of a shell in western China covered in carved signs may be the first evidence of writing in human history. Sometime around 8,600 years ago, someone carved hieroglyphic-like pictographs on each plate of the shell. This predates the previously earliest known writing by some two thousand years.

My turtle is somewhere deep inside his little sculpted domicile. He is lucky; he's been living for decades within sight of downtown Manhattan, yet no little boy or girl has picked him up and decided he'd look better in a cardboard box in the kitchen than in the forest. No car has swerved to hit him as he ambles across the road just to hear the loud pop of his shell. He is, I would guess, somewhere between twenty and fifty years old. Long life is for him not a gift, but a necessity. He may need decades simply to locate enough mates to overcome the odds stacked against replacing himself in the population. Nearly every animal in the forest larger than a few inches will gobble up his babies before their shells develop the hard domed, hinged battlements that have protected him these past decades.

My turtle is a reptile, less primitive than an amphibious frog or salamander but an ancient citizen of the Earth nonetheless. He is scaly, cold-blooded, and his kind breeds by laying hard-shelled eggs on dry land, in nearly all cases depositing them carefully in the ground and then abandoning them. Crocodiles are starkly dinosaurian, snakes are elegant in their spooky tubularity, and lizards are in some ways reminiscent of very primitive mammals. But turtles and tortoises march to a different evolutionary drummer. A tortoise is superficially similar to a turtle, but not at all the same. He has evolved adaptations to a life utterly divorced from water, which has allowed him to invade some of the harshest environments on the planet.

His shell itself is a wonder of nature, a rib cage that natural selection has taken to absurd extremes; both the shoulders and pelvis have been moved inside it. Wearing his skeleton on the outside offers great protection for a slow-moving, defenseless animal. It is a massive organ in itself, making up more than half the weight of a large tortoise, and measuring up to three inches thick in the largest species. Only the odd pancake tortoise has forsaken a rigidly bony carapace for the sake of a flat, pliable one. It lives in rocky outcrops in East Africa, where it can, by virtue of its flat, flexible shell, wedge itself into a rock crevice to make it all but immovable to would-be predators—except humans armed with crowbars and long arms. For the rest of this clan, a hard, domed shell has served well. The carapace can be as domed as a Kaiser helmet or just gently sloped. Seen from above, it can be as circular as a basketball or almost a rectangle. Underneath, the plates of the plastron are jigsawed together in a fairly universal pattern. The carapace and plastron are joined at the midsection by a bony bridge. In some species males have long, projecting front plates, the gulars. These paddle-shaped projections are used to try to flip one another over during the battles of the mating season. If you think tortoises lead quiet lives, you'd change your mind quickly if you attended a mating battle. The ramming of shells can be heard from hundreds of yards away.

The carapace is calcified and overlaid with platelike scales, or scutes, made of keratin, the proteinaceous stuff that makes up your fingernails and toenails. They are built to be tough as nails, too; to withstand falls, the jaws of predators, and the risk of desiccation in the hot sun. As the shell grows, the seams between the scutes reveal evidence of growth in the form of new keratin pushing up around the edges. Over time, these create a ring pattern on each scute that might lead you to think you can tell the animal's age simply by counting the rings. This may work occasionally, but for the most part, you can't. Shell growth is at the mercy of food supply, whether seasons of plenty or of famine, and is not a simple numeric equation.

Underneath the external scutes and underlying the bony shell lies the rest of the skeleton. The neural bones, which are part of the vertebral column, and the pleural bones, which lie along the ribs, are the rafters that make up the interior skeleton supporting the dome of the shell. These bones develop as the tortoise or turtle grows, so the munchable soft-shelled little hatchling will have bony connections in the rib cage. But the ribs of a tortoise don't curve gracefully around from the backbone to meet in the chest or abdomen as ours do. The tips of the ribs insert into the peripheral plates that ring the shell. Attached to the plastron beneath, this dome is as impervious to injury as any skeletal feature in the animal world.

That marvelous shell is our tortoise's last line of defense. The scutes on top, whether drab or brilliantly patterned, are subject to injury and disease, and both they and the bony protection beneath have some ability to regenerate. Many a tortoise has recovered from horrific wounds that left scutes cracked, crushed, or missing. Fire, shell rot, and automobiles are only a few of their biohazards. Veterinarians routinely patch cracked tortoise shells with fiberglass mesh, leaving a bionic tortoise to live many more decades.

A neck of eight vertebrae can be retracted inside the shell to conceal and protect the head, and can be redeployed by strong neck muscles. His sizable lungs (two of them, unlike the single lung of many reptiles) lie just under the upper dome of his carapace, well protected unless run over by a car, in which case respiratory failure is only the first of his problems. Since the rib cage is absolutely fixed, he cannot rely on a mammal's in-and-out movement of the chest to aid in breathing. Instead, his neck, arm, and leg muscles aid in respiration by helping to change air pressure in the lungs with their movements. As he moves his body, his internal organs sink lower in his body cavity, and these create enough negative pressure to make his lungs fill. However, pulling in his head takes up valuable room inside the shell as well, so his lungs must then empty. The flattish, three-chambered heart is not radically different from that of a mammal except in its languid beat. His heart rate, like his basal metabolic rate and even the sex of his offspring, is cued largely by external temperatures. For those species living in temperate climates, like the box turtle, hibernation is a natural part of the animal's life cycle, and of his biological adaptations.

As reptiles, turtles and tortoises are "cold-blooded." But this is a little misleading. Although our little turtle cannot internally control his body temperature, he is not entirely at the mercy of the elements. He is able to maintain his body temperature at equable levels by moving in and out of shade or sunlight. Many tortoises thrive in areas where daytime temperature can reach well over 100 degrees Fahrenheit. But this doesn't mean they spend much time in those temperatures. Desert tortoises of the American Southwest live in the most intense heat of any tortoise on Earth, when during the summer the ground becomes a dry furnace at 120 degrees. But you won't find any self-respecting tortoise on the surface at that time. A few feet underground the temperature remains a comfortable 70–80 degrees, and the humidity far higher than the arid ground above. So the desert tortoises burrow. Safely ensconced in their cooler, moister microenvirons, they weather the daily blowtorch heat, coming above to look for food only at dawn and dusk, and even lying dormant in their burrows during the worst weather. Other tortoises use sand, leaves, and even their own saliva or urine to mitigate the effects of heat. The bladder of a desert tortoise is a vital water storage organ. Innocently pick up a tortoise in the desert, and the stress of capture will make her pee, losing fluids so precious it may jeopardize her life during hot weather. Larger tortoises have the added benefit of sheer size; it takes a long time for a five-hundred-pound animal to heat up, and to cool down once heated by the sun.

A tortoise can survive these kinds of environmental torture not only because it is a tough creature, but because it doesn't have the same metabolic needs as a mammal. Needing only a tiny fraction of the calories and nutrients per day that a similarly sized mammal would, and not needing to waste any calories on metabolic heating or cooling, a turtle or tortoise can survive in places where other animals would starve. Hence the giant tortoises of the Galápagos and Indian Ocean, which exist (or at least used to exist) at population densities that would starve warm-blooded grazers like deer or antelope. Since they do not need to worry much about starvation during lean periods, tortoises also do not need to migrate the way mammals do, allowing them to inhabit hostile places the year round.

Furthermore, many tortoises and turtles can slow their metabolism when times require, such as when food isn't available. Hibernation is one of Nature's most wonderful assets for a cold-blooded animal that sometimes lives in a temperate climate. Cued by decreasing daylight and cooler nighttime temperatures, our tortoise's systems begin to prepare themselves for a long winter nap. Likewise, some species brumate in hot weather, passing the seasonal heat in a state of torpor.

Deep History

Tortoises are often thought of as living fossils, and it's at least partly true. While some of the most ancient-looking giant species are actually quite young in evolutionary terms—

veritable teenagers—the lineage itself is very old. The shell is the thing; its appearance dates back 220 million years to a primitive creature called *Odontochelys,* which lived in southwestern China. *Odontochelys* lived in shallow seas, possessed a skimpy but real shell, and had lots of horny teeth. It's not clear whether *Odontochelys* is telling us that the very earliest chelonians (turtles and tortoises) evolved for life in the water, or if this early fossil indicates the first foray into an aquatic environment by a lineage that had been up to that time exclusively based on land. If the latter is the case, and most experts have long agreed on this second scenario, then the tortoise likely pre-dates the turtle; the latter appeared millions of years later. This later appearance of the turtle would suggest that adaptations to life on land were the forerunners of life in the water. If this sounds backward, it may be because the entire lineage, the sauropsids, emerged from creatures that were already well ensconced on land. Most of the lineage diverged to become the group that gave rise to all the other reptiles, and in a circuitous evolutionary route, the birds.

But why in the world did the shell come to be in the first place? All notions point to an ancestor that had, for reason of protection against predators, developed bony armor of some sort. One theory places the shell's origin in the thickened rib cage of an early progenitor, *Eunotosaurus,* which lived 50 million years earlier than the first tortoise. Other researchers place its origin in a lizardlike critter called *Captorhinus,* whose shoulder girdle had migrated inside the rib cage, much like a modern tortoise. A third and perhaps best candidate is *Anthodon,* a diminutive reptile whose back was covered in armored bone.

Whatever the place of *Odontochelys* or its contemporaries

as the true basal ancestor, one important heir to the evolutionary throne was *Proganochelys,* a beefy yardlong beast that looked a lot like a toothy, tough tortoise. *Proganochelys* lived in Europe and Asia slightly more than 200 million years ago, likely on land but near the water's edge. It looked like an antediluvian version of a modern snapping turtle, but with teeth. Its spiky, bony plates, fused to the rib cage to form a carapace, no doubt gave excellent protection against the myriad predators, dinosaurian and other, of that time. However, unlike a modern turtle or tortoise, it could not retract its large head into its shell. Furthermore, its long tail ended in a spiky club, perhaps added armament against the possibility of capture.

It was only much later that the first land-bound tortoises appeared. Their clan has its deepest ancestry in the same time period in which mammals finally took over the world and our primate ancestors first appeared, at 65 million years ago. For all those intervening eons, turtles had solved the evolutionary dilemma of how to be slow and gently inoffensive but still protected from predators, yet had not moved from their watery home to colonize the dry parts of the globe. But once invaded, the land quickly filled with tortoises. By the Eocene period, some 50 million years ago, there were tortoises across the globe. The earliest known tortoises were probably somewhat similar to the modern Asian forest tortoise (*Manouria emys*). We believe this early Asian ancestor, *Hadrianus,* was the primitive stock from which all living North American tortoises sprang.

If you are accustomed to thinking of tortoises as high-domed reptiles inhabiting arid desert zones, you might not have recognized *Hadrianus.* Today, we think of tortoises as naturally suited to the harsh dry areas of the world, and indeed many species live there. The desert tortoise is highly adapted to the privations of life in its aridly hostile habitat. It avoids the heat and sun like the plague. But it was not always like this. *Hadrianus* lived not in the desert, but in wet tropical forests, like *Manouria* does today. We know nothing about its behavior, but if it was anything like modern *Manouria,* it was a forest tractor, capable of plowing through dense undergrowth and up and down steep hillsides. It laid its eggs in the tropical soil, or perhaps in self-made mounds of debris the way *Manouria* does. It possessed a lower, flatter shell than the high-domed model favored in deserts today. The high dome is presumed to be a better preserver of essential fluids than the flatter shell, allowing a large internal space without exposing a wide body surface to the tropical sun. Having forsaken the water, the tortoise has little need for streamlining; on land, fleeing is not an option. He offers practically no resistance other than camouflage against being found by a predator. Unless secreted below ground, he sits in plain sight, placing full faith in the power of his body armor to protect him from harm.

For many millennia tortoises small and large were a prominent part of the global fauna. Today when we think of giant tortoises, we think of the Galápagos Islands (most of us are unaware that a parallel giant tortoise lives halfway around the world on remote islands in the Indian Ocean). But gigantic tortoises, some of them larger than the giants of today, once roamed all the continents except Antarctica. *Meiolania,* a massively armored eight-foot brute of a tortoise, roamed Australia and nearby islands for 40 million years; it was likely exterminated by the first aboriginal people to reach the islands just a few tens of thousands of years ago. The Caribbean

Islands had their share, as did the land that would become Florida in the United States. The Mediterranean and North Africa had tortoises related to those that roam Africa today, but bigger, and the giant of them all was a massive creature from Asia named, appropriately, *Colossochelys*. This walking boulder lived from India eastward through Indonesia just a few million years ago and had a shell that was more than eight feet long. Related species lived in the Indian Ocean, including islands that today are home to the modern giant Aldabra tortoise.

Who Are the Tortoises?

The tortoises today comprise several dozen species, and since I'm going to be talking about what threatens them throughout the book, it would be a good idea to introduce them. We can divide them several different ways. Scientifically, there are fifteen genera, forty-five species, and more than eighty total varieties, or taxa. We might divide these forty-five species according to the ecological role they fill. Although there is some overlap among categories, it would look like this:

THE LARGE DRY LAND SPECIES. These are the large tortoises of the mainland continents, living in grassland or desert, functioning as herbivores. Some are truly large herbivores, like the spurred tortoise (*Geochelone sulcata*) of central and western Africa, topping out at over one hundred pounds. Leopard tortoises (*Stigmochelys pardalis*) are also African grazers that can reach a huge size in some regions. Others are more modestly sized, like the footlong desert (*Gopherus agassizii*) and Texas (*G. berlandieri*) tortoises of the American Southwest. One of the four North American dry land tortoises, the Bolson's (*G. flavomarginatus*), is considerably larger than the others. The closely related gopher tortoise (*G. polyphemus*) is the only tortoise of the eastern United States. Tortoises that live in dry climates include two of the most cherished and rare of the world's tortoises: the radiated tortoise (*Astrochelys radiata*) and the angonoka or ploughshare tortoise (*A. yniphora*), both from Madagascar and the latter one of the world's rarest animals. Some of these, like the desert, Texas, and Bolson's tortoise, live in truly arid places, spending much of their lives in underground burrows to avoid the hostile heat. Others, like the radiated tortoise, live in arid scrubland areas where there is a rainy season. But all live more or less in open, hot, dry places.

THE SMALL DRY LAND SPECIES. About a third of all tortoises fall into this category, as do many of the species most commonly kept as pets. In northern Europe small tortoises from the Mediterranean have been kept in backyard gardens for centuries. The British cleric and naturalist Gilbert White kept a daily journal for decades of one such tortoise, Timothy, in the 1700s, contributing to his tome *The Natural History of Selborne*. Sailing ships brought exotic spices, foods, and other commodities to the British Isles, and along with them they brought animals to sell. Many, many tortoises from dry, sunny climes ended up in damp, cold British households, coping with the hostile climate for decades in testimony to their hardiness. Today, the Mediterranean tortoises of the genus *Testudo* have been divided by scientists into no fewer than twenty-five species and subspecies, many of which are threatened with extinction.

This group also includes some of the small South African tortoises. South Africa has the greatest tortoise biodiversity in the world, but most of the species have small distributions and must compete with vineyards, farms, and housing subdivisions for living space. In South America there are analogs to the Mediterranean tortoises—small, nondescript species called Chaco tortoises (*Chelonoidis chilensis*), which some researchers consider to be two or even three species. These little brown tortoises have one claim to fame; molecular studies have shown them to be the closest living relatives to the giants of the Galápagos.

Radiated and ploughshare tortoises share their dry land habitat with the tiny spider tortoises *Pyxis arachnoides*. They also spend their lives partly buried in the leaf litter, rarely emerging from cover of brush except during the monsoonal rains. Like the other Malagasy tortoises, these are highly threatened by collection for food and especially for the international pet trade.

The group is rounded out by the beautiful star tortoises (*Geochelone elegans* and *G. platynota*) of the Indian subcontinent and Myanmar, inhabitants of a variety of habitats but most often dry, brushy areas. These are among the world's most stunning living things, a beauty that has come at great cost. In addition to habitat loss and the food markets, they are preyed upon by the thousands to be smuggled to the West as pets.

THE LARGE FOREST SPECIES. We could debate exactly what I mean by "forest," but in general these tortoises live in tropical forests of the world, albeit those including in their realm open grassland and forest fringe. In the Amazon region of South America and also in the wet forests of Southeast Asia live large, lumbering tortoises. These are the yellow-footed tortoises (*Chelonoidis denticulata*) and red-footed tortoises

(*C. carbonaria*) of the New World, and the Asian forest tortoise (*Manouria emys*) of the Old World. All three function in a similar role. They roam the wet forest floor, searching for fallen fruit and scavenging plant products rather than subsisting on the dry weeds and grasses that make up the diet of their dry land counterparts.

THE SMALL FOREST SPECIES. In many of the same tropical forests where large tortoises live, there are smaller species as well. These also tend to be secretive forest-floor fruit-eaters, and they also tend to be high on the menu of many people living in or near tropical forests. The beautiful impressed tortoise, *M. impressa,* is one such shy species. It spends much of its life under leaf litter high in the mountains of Southeast Asia from Myanmar eastward to Malaysia. In some of the same forests (as well as drier ones), one can find a second forest species, the common and widespread elongated tortoise (*Indotestudo elongata*).

In equatorial Africa, hingeback tortoises of the genus *Kinixys* are forest-floor dwellers—shy creatures notable mainly by the posterior hinge that allows them to shut their shells tightly. They are not brilliantly patterned like the starred tortoises, but most species have a subtle beauty of warm colors and a timid demeanor. Like all the tortoises of wet forests, their main foods are from the forest floor—fruits, fungi, slugs, snails, and the like—not the grasses and weedy plants of open country. One species, Bell's hingeback, is ubiquitous enough to be found in many habitats, from wet forest to dry savannah and everything in between.

The flat-tailed spider tortoise, *Pyxis planicauda,* is an inhabitant of much lusher forests than its spider tortoise relative, *P. arachnoides.* In Kirindy forest in Madagascar, one of its last strongholds, this tiny tortoise brumates for substantial portions of the year and is found out and about only during the rainy season.

THE ROCK SPECIALISTS. There are a few tortoise taxa that do not fall neatly into any of the four groups I have mentioned. They live in grasslands in the broadest sense; but really, they inhabit rocky areas that are scattered across the sea of grass. In a limited area of East Africa, the pancake tortoise (*Malacochersus tornieri*) lives in the *kopjes,* or rock outcroppings, that dot places like the Serengeti plains in Tanzania. They come out to forage in the daytime but spend much of their lives wedged deep inside crevices.

In South Africa, the smallest tortoises in the world, while living in grassy areas, spend much of their time hiding under rocks as well. The padlopers (genus *Homopus*), all four to five inches of them, are by virtue of size, or lack of it, potential prey for many animals. They avoid such dangers by being highly secretive.

THE ISLAND GIANTS. One group deserves to be considered separately from all the others, although the group's general ecology is not so different from those large species living in dry lands. These are the giants of the East Pacific Galápagos archipelago and of the Indian Ocean atoll of Aldabra. Their origins, history of maltreatment by people, and current status are the focus of Chapter 5. As island forms highly adapted to the isolated places they have found themselves for the last few million years, they are unique for much more than their vast bulk. In addition to the living two genera and at least two species, there is one recently extinct genus that includes at least five species. Of at least fourteen total varieties of Galápagos tortoises alone, perhaps three are extinct today, with a fourth reduced to one pathetically lone representative of his former clan.

There are a number of species listed in the back of this book that defy easy categorization. They may be so ecologically versatile that shoehorning them into a particular habitat doesn't work; the elongated tortoise of Asia, for instance, occurs in both tropical rain forests and dry scrubby open areas. Or they may have a wide span of sizes in different parts of their range. The leopard tortoise can be a fifty-pound monster in the Western Cape region of South Africa or a modestly sized ten-pounder on the Serengeti plains of Tanzania. But my point is that nearly every habitat type on Earth has a tortoise. If an area doesn't, it tends to have some other form of turtle that is highly adapted for life divorced from water.

What Is a Species?

There is another thorny issue embedded in the battle to save endangered species, including the rare tortoises. This is confusion over classification itself. To protect and save, we must understand. To understand, we must set some ground rules. Such as, what is the basic biological unit that we seek to save? Defining a species can be as slippery a topic as exists in the life sciences. Species are not neat pigeonholes. They're dynamic entities, fluctuating in time and space. Historically we define a species based on appearance: a sparrow with a gray head is easy to distinguish from one with a brown head. But what of the intermediate forms with grayish brown heads? These can be ignored, or they can be considered a dubious third species, or the two original forms can be further subcategorized by labeling them as subspecies. In the case of tortoises, forty-odd species have been subdivided into a total of about one hundred subspecies.

The Swedish botanist Carolus Linnaeus devised the modern science of classification, so we owe him our thanks and our scorn. Linnaeus lumped together animals and plants based on physical characteristics that he could easily see. You could do the same with cars. Sedans, minivans, SUVs, and pickups all represent distinct categories. There's no confusing a BMW convertible with a Ford pickup. You could then sort your categories into a wide variety of lower-level categories: subcompacts, compacts, and so on. Each of these could in turn be subdivided based on other descriptive features such as color and style (five-speed or automatic). An overwhelming variety of vehicles can be quickly sorted as you're shopping on the internet for the car of your dreams.

But sorting species is not like sorting car models. Cars are designed and built to humans' specifications, and can be tweaked this way or that on the whim of the manufacturer or the influence of the buying public. Animal species change too slowly and incrementally for us to see it in our puny eighty-year life spans. A species doesn't care whether or not we can identify it as such; its boundaries are determined by the animals' willingness to mate with animals from other, similar species. The human tendency to treat species as utterly distinct has led to terrible confusion about orderliness and variety in Nature.

Take lions and tigers. They are two very distinct cats, the regal lion with his mane, the majestic tiger with its black stripes and orange fur. But these differences are skin deep. If you put a lion and tiger of opposite sexes together in a zoo enclosure, a male lion and female tiger will produce a hybrid cub called a liger. Ligers are huge compared to either lions or tigers. Female ligers can mate with either male lions or male tigers and produce fertile, healthy offspring. In the one place in Nature where the two species co-exist—the Gir Forest in western India—lions and tigers have never been seen to mate with each other. So are lions and tigers two distinct species? They are, because in nature the species do not cross-mate. There are many such examples of animals that do not ever meet in nature because they live far apart or occupy different niches in the same habitat, but they would mate with each other promptly if placed in the same cage or pond.

Defining a species is therefore far from easy. With so many of the world's animals in danger of imminent extinction, scientists are desperate to measure the remaining populations. But when one species of tortoise consists of 1,500 remaining animals and is further subdivided into three subspecies of 500 each, the results have critical consequences for conservation efforts. So scientists argue endlessly about ethereal concepts like species because they bear directly on how many species exist, which in turn decides just how many endangered animals need our help.

Scientists who classify species or subspecies have a new arsenal of research tools, thanks to the development of sophisticated genetic techniques in the past decade. We used to rely entirely on the outward appearance of an animal, plus some aspects of skull or skeleton, combined with its geographic distribution and preferred habitat to decide whether one species should be broken into two. For instance, the Mediterranean tortoise *(Testudo graeca),* so often kept as a backyard pet in Europe, occurs all across southern Europe into North Africa, and eastward across the Mediterranean all the way to

Iran and Turkey and southward into Israel. Across this huge range it differs in size, color, and shell pattern. It occurs in bushy areas, rocky areas, and barren places, from the furnace of a Moroccan summer to the bitter cold of a southern Russian winter. Some researchers feel these differences warrant splitting *T. graeca* into as many as sixteen separate species. But molecular genetic studies do not support this fragmentation. The genetic distinctions among the many populations are not sufficient to indicate that any of the groups has gone off on its own evolutionary path. So instead of sixteen species, most authorities consider *Testudo graeca* to be one wide-ranging species with sixteen subspecies, or races. By definition these subspecies do not overlap or co-exist in the same spot. If two subspecies were found living side by side without cross-mating, it would strongly suggest that they had already diverged enough to be considered separate species, hence their lack of interest in mating with one another.

The effort to decide which species should be merged into others and which should be pulled out of existing species and given their own name is hotly contested in both scientific and political arenas. Scientists have differing views of what constitutes a species, so even genetic data are not always accepted by the field at large without lengthy philosophical debates. Each newly named species also represents a new entity worthy of legislative protection. This may create political battles over everything from land development practices—that housing subdivision can't happen without wiping out an entire species, rather than just a tiny fragment of one—to arguments over which species deserve the most funding for conservation programs.

In recent years the study of genetics has superseded traditional rationales for species categories based purely on external appearance, but unfortunately most of the world's pet owners don't pay attention to genetic studies. Those tortoise species that are most vulnerable to poaching for the pet trade have been creatively labeled and relabeled by reptile dealers and breeders who are more concerned with marketing and sales than biological reality. Thus a local population of the *Testudo graeca* that was identical but paler in color and pattern became the "golden" Greek tortoise in hopes of attracting buyers to a "new" tortoise. Nice try, except genetic studies clearly show that there is no basis for separating the less-patterned animals from the species at large. Such mislabeling not only contributes to biological confusion, it hampers conservation efforts.

The Tortoise's Niche

Since the premise of this book is that it's imperative that we act quickly to protect remaining wild tortoises, there is an onus on me to convince you that these are animals worth saving. They are on two counts. First is the aforementioned set of astounding adaptations that make these shelled creatures among the most remarkable organisms ever to plod slowly across the Earth. The second reason is the role tortoises play in modern ecosystems. Tortoises may not seem the most vital members of the natural world. But that is not true. They are primary consumers that graze or browse as hoofed animals do, though usually on a much smaller scale. Tortoises come in a variety of shapes and sizes, but all are primarily herbivores. They range from little vegetarians that skulk under rocks and in burrows, coming out in the cool and safety of dawn or dusk, to hulking tanks that chow down dozens of pounds of fodder a day, converting an amazing percentage of it into their own biomass.

Today, there are tortoise species as small as pebbles and as large as boulders. There are tortoises with shells so thick and durable that a fall from a cliff might not be fatal, and other species with shells so pliable that the turtle can wedge itself into a crack in a rock so tightly only a crowbar can extract it. The little species include the speckled padloper, the smallest tortoise on Earth. Little more than a tennis ball with legs, it lives among and under the rocky hillsides of its native South Africa (where "padloper" means "tortoise"). Like most tortoises that live in hot sunny climes, the padloper spends much of its life out of sight, avoiding the killing rays of the sun. At the other extreme of size are the island giants as well as some not much smaller species in Africa and Asia.

The large tortoises are often called reptilian cows, and for good reason. They are enormous grazers that stand in place in pasturelike fields all day, munching grasses and herbs. They are sociable, or at least they gather in great numbers to feed in the best places, giving them a herdlike appearance. Like cows, theirs is an existence based on easy access to widespread, abundant, high-fiber foods. And like cows, they have been treated by people over the centuries as a ready source of meat on the hoof.

But there is one fundamental difference between a cow and a giant tortoise: their metabolism. A cow eats from dawn to dusk because grass and weeds offer relatively low nutrition and few calories, compounded by the need to break down all that fiber. The tortoise can live on the same diet, but because it is not a mammal it can go months without a constant source of good food. Long after the cow would have died of starvation, the tortoise grazing alongside would be happily taking whatever food it could find in impoverished circumstances. This is why bleak offshore islands can support stunningly high densities of tortoises. They just don't need as much food as a warm-blooded herbivore of the same size. They can survive in places where drinking water is nearly nonexistent. They are, in a word, tough.

The tortoise therefore occupies a niche in the ecosystem as a low-metabolism grazer. At one time in Earth's history, before humans stamped out much of our megafaunal diversity, they were mega-grazers in ecosystems from Florida to Australia plodding alongside antelope, elephants, and ground sloths. As beautiful and strange as modern Madagascar's tortoise fauna are, there were two additional species—giant ones at that—that were driven into extinction only recently when humans discovered them. Ancient tortoises may have lived in densities that approached the herds of hoofed animals on the same plains. Their massive shells were an effective bulwark against the toothy predators of the day. But when humans arrived, continent by continent, they found tortoises to be easy prey, and tortoises rapidly disappeared.

One aspect of eating plants is that the plant-eater finds its guts full of seeds. In all ecosystems, seed dispersers play a crucial role, and many plants and trees have evolved ways to persuade plant-eaters to pluck their fruits. Since seeds are the reproductive output of a tree, it's vitally important for the tree to get its progeny far away from it, out of its shade and the competition for scarce resources. This is where seed dispersers play a role. When an animal eats fruit, the seeds are excreted in the droppings, and if the plant-eater has walked a great distance since consuming the plant, the seeds are dispersed far and wide. The bright colors, sweet taste, and soft texture of fully ripe fruit are believed to have evolved as signal flags that tell a fruit-eater that it's time to pluck those fruit. Herbivorous animals fulfill this mission the world over. There are trees that have evolved specific seed hardness to ensure that the local fruit-eating mammals can't crack their shells, forcing the consumers into a disperser role. The ability of a tortoise to disperse seeds is limited by its own mobility. Some species can plod long distances in the course of a few days, however, so large plant-eating tortoises can shape an entire ecosystem.

In addition to their main role as grazing herbivores, in some times and places tortoises can be predators and, if not exactly ruthless killers, omnivores that will chomp down anything that moves within their slow reach. So the forest tortoises, which comprise dozens of species in most tropical regions of the world, occupy the niche of forest-floor vacuum. They chase down snails and slugs in slow motion and eat baby mice and other tiny creatures.

Tortoises can also be scavengers. One morning I walked into my backyard and came upon my group of Asian forest tortoises huddled around a large dead rat, which they were ripping apart like lions tearing into a zebra carcass on the Serengeti. These tortoises are extremely fond of meat, and most species eat not only meat but bones, insects, and whatever other little scraps of animal protein and fat they can get their beaks on.

So the ecological role of the tortoise is that of a small (with notably colossal exceptions) herbivore. But its other ecological role is as prey for many other species. In the Amazon, jaguars and other smaller carnivores flip tortoises like beach balls, trying to get purchase with their sizable teeth so they can tear into their innards. Tigers and leopards no doubt do the same in Asia. The tortoise is well protected, but if the predator is strong enough, it is a can of meat waiting to be opened.

Recently, researchers in central Africa studying chimpanzees discovered that the apes forage for tortoises as well. Lacking the teeth and claws needed to penetrate the shell, a chimpanzee that comes across one turns to an even better solution. He uses his hands to smash the tortoise. Researchers aren't sure yet whether this is done by smashing the tortoise against the ground like a beer can, or by employing rocks as hammers to bash open the shells. Either way, the apes crush their prey and get at the sweet meat inside.

Fully appreciating the uniqueness of a tortoise, however, requires a discussion not only of its amazing anatomical adaptations, but also of the extraordinary aspects of its biology, such as its extreme life span. So I turn now to an examination of the role that a long, slow existence plays in the evolutionary success story that is the modern tortoise.

2

LIVE LONG AND PROSPER

In the autumn of 1835, as the H.M.S. *Beagle* hoisted sail and headed away from the Galápagos Islands bound across the Pacific toward Tahiti, she carried many more passengers than she had arrived with five weeks earlier. Captain Robert Fitzroy's crew had also collected thirty huge tortoises to store alive onboard as food. Ship naturalist Charles Darwin had collected three very young tortoises and decided to try to bring them back to England alive. Darwin and his assistants named their new pets Tom, Dick, and Harry. Tom had been collected on San Cristóbal Island, Dick on Floreana, and Harry on Santiago Island. They lived in a box in the dank below-decks cabins of the sailing ship for another year before the *Beagle* docked at Falmouth, home for good. The three tortoises took up residence at Darwin's London home, the only living specimens from an expedition that returned with thousands of dead and pickled new creatures.

From there the story grows murky. All we can be sure of is that a century and a half later, many people believed that an ancient tortoise living in a zoo in Australia was the same Harry that Darwin had brought home from Santiago. Harry (later renamed Harriet upon proper sex identification) had, according to some accounts, been brought to Australia from England in 1860 by John Clements Wickham, an officer on the *Beagle* during her historic voyage and a friend of Darwin's thereafter. Seeing that the three little tortoises were not coping well with English winters, Wickham offered to take them with him to a more tortoise-friendly climate, and Darwin agreed. Two of the three died during the course of the twentieth century, but Harriet lumbered on, surviving to the age of approximately 176 (presuming that she was about five years old when collected by Darwin in 1835) until a peaceful death in 2006.

There are doubters who don't believe Harriet to be the same tortoise that Darwin brought back, and they may be right. But if Harriet was not the same Harry who was once Darwin's household pet, it wouldn't be because her great age of 176 is implausible. It is far older than tortoises live in the wild; there are just too many risks to life and limb in the natural world. But in the comfort of captivity it is a plausible ripe old age, rather like a person who has survived to 100.

There are plenty of examples like Harriet. Traveling the South Pacific, Captain Cook presented the king of Tonga with a tortoise (a radiated tortoise named Tu'i Malila with a beautiful starburst pattern on the shell) as a gift in 1777. In 1966 Tongans mourned its death at an alleged age of more than 180. Some doubt the veracity of the reported age. But because radiated tortoises are rare and naturally occur only in Madagascar, it seems unlikely that a second radiated tortoise would have arrived to replace the first. Addwaitya, a giant Aldabra tortoise, died recently in the Alipur Zoo in Calcutta, India. He was given to the zoo by a British colonial officer in 1875, so at his death in 2006 he would have been at least 130. Esmeralda, an enormous male Aldabra tortoise living on Bird Island in the Seychelles, is alleged to be more than 200, making him a candidate for both largest (over eight hundred pounds) and oldest chelonian known.

Mammals, though they may live relatively long lives, are most notable for their drawn-out life cycle. The sheer length of their lives is not as impressive as the amount of time it takes them to grow up. Of course there is a wide spectrum here too. A mouse is born naked and blind after a pregnancy of twenty days, but within two weeks it's running around playing with its siblings, and a few weeks later it is nearly ready to have its own offspring. By the age of two years, the mouse is elderly. A rhesus monkey, on the other hand, is born after a five-month pregnancy, takes three to five years to reach sexual maturity, and lives up to twenty-five years.

One day in the summer of 1997, Jeanne Calment decided she had lived long enough. She stopped eating, and over a period of a few weeks her health declined and she died. Madame Calment had been born in France in 1875, and so was 122 years old at the time of her demise. She was the oldest known person ever to have lived whose birth date could be authenticated. Calment had outlived her children and some of her grandchildren. She took up fencing at age 85, and rode a bicycle at 100.

Humans certainly live longer lives than a great many creatures on planet Earth, including most mammals. We live longer than any of our closest relatives, the great apes. The oldest record for an ape is at least the sixty-six years that a captive chimpanzee named Gregoire lived in a sanctuary in

Africa. (Cheetah of *Tarzan* fame was recently dethroned when it was revealed that he was decades younger than his purported age.)

Some mammals live far longer than any primate. Whales can live well past the century mark, and perhaps even more than two centuries. In 2007 a fifty-ton bowhead whale was caught off the coast of Alaska, and when it was cut open, a fragment of harpoon head was found imbedded in its neck. That would be no great shock, as many whales carry the scars and evidence of past hunting. But this harpoon head, dug from deep scar tissue, had been manufactured in the whaling town of New Bedford, Massachusetts, in about 1890. The whale had been hunted sometime around the time that Rutherford B. Hayes was the U.S. president and lived to the venerable age of at least 130 before it was killed in a subsequent hunt. Further investigation turned up a whole series of harpoon tips from recently killed bowhead whales, some of stone and at least one of ivory. One of those whales was estimated to be 211 years old at its time of death. A few other mammals, notably elephants, also live more than sixty years, but beyond the primates, cetaceans (whales and dolphins), and proboscideans (elephants), extreme longevity among mammals is rare.

The causes and consequences of long life have fascinated scientists for centuries and spawned entire research fields. Considered in light of the knowledge Darwin has bequeathed us, long life is not a gift, but an evolutionary strategy designed to maximize reproductive output. Those animals that live long lives also tend to take a very long time to pass through life's stages. There are several common corollaries of long life in the animal kingdom: large body size, slow reproductive rate, a social nature, and a large, complex brain. This combination of brain size and social complexity may account for longevity and a protracted life cycle.

Why Do They Live So Long?

Everything ages, right? Not exactly. Bristlecone pines cling stubbornly to windswept ridges in the western United States. They don't age in any conventional sense. They apparently experience no buildup of harmful mutations like the rest of us during their five-thousand-year lifespan. Bristlecones are just as able to transport water from their roots to their stems at age three thousand as they were as young plants. Seeds produced by the "Methuselah" bristlecone (approximately 4,715 years old at its last birthday) are just as likely to germinate as are seeds from hundred-year-old teenagers and fifteen-hundred-year-old midlifers, and pollen is just as viable from ancient trees as from young ones.

Senescence is the process of aging. Tortoises and turtles do senesce, but as every child knows, they can live a very long time. Although their average lifespan in the rigors of the wild is not necessarily so extreme, they are certainly among the longest-lived of all vertebrates. Captive animals don't have to cope with the privations of food and water, or danger from predators, and so they often live far longer than their wild counterparts. Most tortoise species are known to live at least twenty-five years in the wild, and half the species in a fairly random sample made by researchers some years ago were found to live more than a half-century. Detailed information is hard to come by owing to the difficulty of accurately determining the age of a tortoise that was alive for decades before the studies began. The major causes of death, not including human factors, are predation and extreme environmental hazards such as deep freezes and droughts.

We take the extreme longevity of tortoises and turtles for granted. Instead, we should ask why they live so long. We can assume that evolution would not have favored extreme longevity unless there were some critical benefit to it. What is the evolutionary advantage to long life? One corollary of long life is delayed maturation. Many tortoises do not start reproducing until their second decade; even some of the common aquatic turtles living in our local pond must be close to ten before they reproduce. The giant tortoises don't reach sexual maturity until they are past twenty years old.

You might think that any animal would want to reproduce early and often. Darwinian theory would seem to support this. But the goal is always "how best to leave many offspring *that survive to reproduce themselves.*" Having fistfuls of offspring won't get you anywhere in the evolutionary race if those babies don't survive. We believe delayed maturity and long life evolved for the simplest Darwinian reason—those individuals that long ago deferred breeding (presumably due to a mutation that caused delayed maturity) ended up leaving more surviving offspring—having higher reproductive fitness in the biological sense—and that pattern became genetically entrenched in the species. Since most turtles and tortoises show this pattern, there is every reason to think that it has worked for many millions of years.

There are rewards for delaying the age of first reproduction. Offspring may be hatched at a larger size or a more advanced state of maturity. Conversely, a larger number of offspring may be produced when breeding finally begins. Late maturity may also reduce the mortality risk to adults, since mating and egg laying often exposes tortoises to predators. Longevity, plus a delayed maturity and late age of first reproduction combined with low annual offspring output, provides an evolutionary formula that has worked well. But this same combination of traits can be deadly when confronted with rapid-fire humanmade disasters, like habitat destruction or overcollecting. Absent human interference, the suite of evolved factors that characterize modern tortoises would have ensured their survival on Earth for eons to come.

Eternally Fertile

It is not only sheer longevity that benefits tortoises and turtles. It is also the power of their unceasing fertility. You cannot separate longevity from reproduction when talking about animals that reproduce throughout their long lives. Some animals produce masses of eggs in a short lifespan, casting pearls to the wind in hopes that a few may take hold. Other animals produce very few offspring in a lifetime, but compensate with many years of nurturing care to ensure that their offspring survive to maturity. Tortoises have opted for a third reproductive path. They don't produce huge numbers of eggs. Except for a couple of species, they don't care for the eggs once laid; the female digs a hole with her hind legs, plops them in, kicks some dirt over them, and walks off, never to think of them again. But they continue laying eggs year after year after year. What they may lack in annual success, they make up for in persistence.

Turtles and tortoises that live in temperate climates are seasonal breeders, their physiological cycles mapping those of the bleak winters, forgiving springs, and blazing summers. Others cope with the more subtle but equally persuasive cycles of a tropical forest. Even in a rain forest, there are strong seasons; they just don't correspond to our notion of what a season is. Some fruit trees blossom in May; others standing right next to them may blossom in October. Each of these seasons exerts a strong pressure on the reproductive timing of wild tortoises. Males in temperate climates produce enough sperm to last them through hibernation for use the following spring. Long-term sperm storage is the norm for both sexes, but for different reasons. Males need to keep sperm in cold storage to be ready on a moment's notice when the spring thaw comes. Females retain sperm both to avoid laying eggs when the cold weather is coming and also to avoid the risk of failing to find a mate. Just as the rest of the creature's makeup is about the long term, a female cannot determine the likelihood of bumping into a male in a given year. Sperm storage allows to her carry active sperm in her oviducts for up to several years and to lay eggs repeatedly without having to encounter a male each time. Studies have shown that a male and female box turtle can be only a few yards apart and not find each other for mating. Likewise, a desert tortoise may not bump into a male simply because she spends a high percentage of her life in her underground burrow. So sperm storage is another testament to the tortoise's live long, live slowly strategy for success.

Environmental cues, whether temperature, rainfall, humidity, or increasing daylight, bring about hormonal changes that prompt male tortoises to go into battle against each other. In some species combat is long and hard, with possibly lethal consequences. The males may use specialized lengthened gular plates of their plastron, just under the chin, to flip over rivals. In a hot sunny climate, being flipped and unable to right yourself is fatal. Males of some otherwise placid tortoise species are almost maniacally aggressive toward one another, chasing and ramming and battering each other in a hormonally poisoned game of king of the hill. As Darwin originally predicted, in those species in which males fight over females, males tend to be larger than females. Size matters. In other species, females may be larger in order to carry more eggs.

At all other moments of their lives, tortoises are among the quietest creatures on Earth. If you pick up and roughly handle a wild tortoise, you may get a hiss as he pulls his head and neck into his shell and tells you he's not happy about it. But mating season brings out a different tortoise altogether. A male tortoise mates vociferously, and often. He will follow an uninterested female until she relents, then mount her and grunt, cluck, or chirp noisily as he mates or tries to mate with her for hours. In some species, a blissful groan can be heard a hundred meters away.

All the while, the male is trying to insert his penis into the female's cloaca. This is no easy task, since she has a hard shell covering much of the cloacal region of her hindquarters and can do her best to keep him out of it. So the male tries to stay nearly vertical behind her, gripping her with his claws, in order to slip his organ underneath her carapace and inside her. A strongly concave plastron helps the male stay mounted on the female's domed carapace. If he is much smaller than she is, he may flip backward in his attempts to mount her. Other than occasional eversion while soaking in warm water, this is about the only time a male tortoise's penis is on display. It's a lengthy soft pink thing that typically ends in what looks like a soft spatula.

After sex, the female is on her own to find a spot to lay her eggs. She may nest every year, or several times a year, or only in alternate years. Some tortoises lay so many eggs at a time that a hen would gasp; others barely produce an egg a year. The Asian forest tortoise and the impressed tortoise of the genus *Manouria* are the only tortoises that build genuine nests and provide maternal care for their eggs; a female drags leaves and debris from the forest floor together into a two-foot-high mound not unlike the one an alligator makes. She mounts the hill she has constructed, digs a hole in the center, and drops in her eggs—up to fifty in the case of the Asian forest tortoise—then sits atop the nest. She will aggressively drive away any and all intruders, from lizards to dogs to humans. After a few weeks she gives up all attempts at protecting the nest and walks off, and the eggs are as much at the mercy of the elements as any other reptile's would be.

Contrast this with the nesting behavior and egg output of the tiny flat-tailed spider tortoise of Madagascar, which lays only one or two eggs at a time, or the pancake tortoise of eastern Africa, which scrapes away a bit of dirt and lays one egg every so often during the year. Or the bizarrely outsized egg of some of the smallest tortoises, such as the speckled padloper, whose single egg may have a diameter almost one-quarter as large as the mother's total size. Apart from these extremes, however, most tortoises produce a dozen to two dozen eggs per year. These starkly different egg-laying strategies evolved over millions of years and worked well in the ab-

sence of severe predation threats, such as they face from human poachers today.

In higher animals, the sex of an offspring is determined by
a pair of sex chromosomes. In humans, this is the twenty-
third pair. In some apes it is the twenty-fourth pair, and in
dogs it is the thirty-ninth pair. A female offspring has inher-
ited an X chromosome, a male baby a Y. Because the odds
of inheriting either male or female chromosome is fifty-fifty,
the ratio of males to females in such animals is one to one.
But most reptiles, including tortoises, don't have sex chromosomes. Sex in reptiles is determined not by genes but by the
temperature of the environment surrounding the eggs. This
temperature-dependent sex determination, or TSD, was discovered only in the 1970s. In turtles and tortoises, eggs above
a certain temperature, usually around thirty degrees Celsius
(roughly eighty degrees Fahrenheit), tend to develop female
embryos. Eggs below that temperature tend to develop males.
(The pattern is exactly the opposite in alligators and crocodiles, in which warm incubation temperature produces
males.) If the average temperature during incubation is right
around the key threshold temperature for the species, an even
mix of males and females is produced. The condition is nu-
anced—in some species it seems that there may be some genetic sex determination in addition to environmental sex determination, and the exact mechanism for TSD is not fully
understood. But overall this is a very different way to reproduce than what we see in warm-blooded animals.

TSD means that the mother tortoise can to some extent
influence the sex ratio of her clutch of eggs. If she lays her
clutch in the shade of a tree, she can produce a preponder-

ance of male babies. If she lays in a sunny spot of open beach, she can produce mainly females. We presume this odd feature of reptilian biology evolved as a way for females to vary the sex ratio of their clutches to maximize their reproductive success, although the precise advantages of male versus female babies may vary depending on circumstances. For those species in which larger females lay more and larger eggs that produce larger babies with higher survival rates, there would be an obvious benefit to producing more female babies. They would eventually yield more grand-offspring. This scenario wouldn't apply to species in which males are larger than females, however, which is the case for many tortoise species.

Whatever the right explanation for TSD in tortoises, it has major implications for the future of many species. As global warming brings about a minor rise in Earth's average temperatures, we can predict that tortoises will begin producing more female-skewed clutches. In the short term this might actually benefit the species. After all, since males provide no paternal care of offspring, female tortoises don't need males around except to fertilize them every year or so. But in the long term, global climate change would result in so few males that females might have trouble finding mates, spelling doom for many tortoise species.

We like to think that humans along with the other placental mammals have evolved the most sophisticated form of reproduction on the planet. We conceive our offspring through internal fertilization, carry the embryo in a warm, safe liquid uterus nourished by the mother-to-fetus spigot of a placenta, and then deliver a baby that has most of its neurons working, if not fully up and running. But there are other ways to achieve reproductive success, and other animals have adaptations that perhaps we should envy. One such adaptation is diapause. In diapause, a developing embryo stops its course of growth and remains dormant for some period, sometimes months, before resuming its development. Marsupials like kangaroos practice diapause, and we think it is an adaptation to an unpredictable environment. If a drought arrives and times are tough, the kangaroo mother can somehow stop her embryo from developing further, then resume the embryo's development later when the rains have come and food is once again plentiful.

Diapause in tortoises usually happens after an egg has been laid but before it hatches, although in some turtles the mother holds an undeveloped egg in her oviduct for months until ideal egg-laying conditions arrive. At some point during incubation, the egg stops developing in response to the surrounding environment, often in response to cooler temperatures. Since some tortoise species lay eggs that take many months— sometimes close to a year—to incubate, the growing embryo must survive through various climatic conditions as seasons change and bring cold or hot and wet or dry conditions. In cooler climates this allows a tiny tortoise embryo to spend the winter inside its shell, safely underground, awaiting the warmer spring temperatures. The diapause switch can also be turned on late in embryonic development. In some tortoises, incubation can only proceed following either early or late diapause, which causes breeders working with captive endangered species many headaches. A perfectly healthy clutch of eggs will simply not hatch unless they are cooled down for a period—the antithesis of our expectations about what eggs need. The eggs of the beautiful radiated and spider tortoises of Madagascar are noteworthy examples of this.

Diapause helps the embryonic little tortoise on its way, but the clutch of eggs still faces grave risks. A high percentage of nests—in some studies over 80 percent—are dug up by predators and their eggs quickly eaten. Once hatched, the newborn tortoise contends with sundry threats even without human depredations. Mortality in the first year out of the egg may exceed 90 percent of the clutch. Small carnivores eat them. Birds feast on them. Even frogs, crabs, and rodents eat them. Most of the remaining 10 percent die before their second birthdays. Without a thick shell to protect them, newborns are just hors d'oeuvres for a host of animals. The evolved strategy for babies is to stay hidden and inconspicuous for the early years of life, until time and shell growth provide them with a decent shield. Even baby giant tortoises of Aldabra and the Galápagos are nearly invisible in the wild, staying out of sight until they have reached a less predator-friendly size.

The Fountain of Age

Fertility, reproduction and aging itself remain poorly understood aspects of many long-lived animals because it takes so many decades to compile meaningful information on longevity and reproduction. Only recently did we learn that whales can live as long as giant tortoises, but their marine realm does not lend itself to detailed research that might help explain why. Gerontology—the science of aging and longevity—finds turtles and tortoises as fascinating and important as any animal. This is not only because they live so long. It's because female tortoises and turtles remain steadily fertile and fecund until the end of their lives. They don't lay a hundred thousand eggs in a clutch like a frog, or nurture and protect their young like a chimpanzee. But once they begin reproducing at a relatively late age, they not only do not stop, they never slow down.

Only a few field studies have followed turtle or tortoise life histories and reproduction long enough to learn something meaningful about them. Justin Congdon studied the breeding and survival patterns of Blanding's turtles in Michigan for more than four decades. Blanding's turtles are tough creatures that live in the north central United States, enduring long, harsh winters by spending a good part of their lives hibernating. They spend their summers in and around ponds. Congdon marked the turtles and then trapped them each spring and summer to monitor their growth, longevity, and reproduction. He found that the annual mortality of adults was low; nearly all breeders survived from one year to the next. This was surprising. We might expect that occasional severe winters would kill off some turtles in hibernation, or that their ramblings onshore would leave them vulnerable to foxes, raccoons, and a host of other predators from which they are safe while in the water.

For nearly every other animal species that we know about, a female's fertility declines as she ages. In humans, female fertility ceases at about age forty-five or fifty, when women undergo menopause. We don't see such a well-defined end to fertility in most other mammals, and we know very little about fertility decline in reptiles. But what Congdon discovered in his turtles was a major surprise. The number of eggs that a female turtle laid every spring and the number that hatched did not decrease at all as the females grew older. In fact, females estimated to be over fifty years old were actually *more* fertile than females that were only twenty years old. This tells us that at the very least, a female's reproductive system—her ability to conceive year after year and her ovaries and egg-formation machinery—stays in excellent shape at a stage of life at which those of all other animals are in steep decline. If this pattern turns out to be widespread among tortoises and turtles, it will be a powerful statement of the value of the live slow, live long way of life.

Life-history theories try to parse out the reasons that some animals, including humans, mature so slowly and then live so long. We believe that humans take so long to grow up because so much of what it means to be human must be learned. Because we are born helpless and are by nature intensely social, a lengthy period for learning and socialization has been built into the human lifespan. Tortoises, with their extreme aging trajectory, beg life-history questions. There are two prominent theories about reproductive aging in egg-laying reptiles. The relative reproductive rate theory holds that animals that begin to reproduce later in life will have a higher reproductive output than younger ones, perhaps through larger egg clutches, larger egg size, or by laying eggs more often. A contrasting hypothesis—the senescence theory—argues that older individuals' fertility will decline. The latter is the most widely applied notion, supported by the observation that in nearly all animals and plants, younger individuals are more fertile than old ones. Tortoises may be one of the rare exceptions. In many turtles—which do not typically live as long as tortoises—females tend to grow larger over time, and the added body size allows them to produce larger clutches of eggs. Justin Congdon's work with Blanding's turtles showed that two key determinants of ultimate adult body size are how quickly the turtle grows up, and what age it eventually reaches sexual maturity.

Turtles that reached sexual maturity later in his study tended to ultimately grow larger. Contrary to conventional wisdom, his Blanding's turtles did not continue to grow throughout life. This means that the size of an adult turtle can be quite misleading; it may not mean the turtle is very old, only that it had a long, slow period of growth and development before reaching maturity, when it basically stopped growing.

The senescence theory doesn't hold up well for Blanding's turtles. In Congdon's study, the oldest females, some more than fifty years old, reproduced more often and more successfully than younger females. It seems likely that this turtle's fountain of age applies in other turtle and tortoise species as well; in fact there is precious little evidence that females of any tortoise species lose fertility as they age.

But this protracted fertility does not mean that you will regularly encounter centenarian tortoises. People who find desert tortoises roaming the southwestern U.S. deserts assume them to be ancient when most often they are not. Studies have found that the oldest individuals in a desert tortoise population are perhaps fifty, and others often substantially less. An old tortoise found crossing a desert highway may in fact be only in its thirties, and nearing the end of its natural lifespan. Life is harsh, and risks to life and limb are many.

Natural selection molded the tortoise to be capable of surviving decades in very harsh conditions, so when placed in a comfy zoo, the animal is so over-engineered that it may live even decades longer. In fact, tortoises are so adapted to a harsh life of regular food shortages and poor food quality that when presented with an abundance of "high quality" food, it can kill them. A dry land tortoise in its natural environment, if it survived its early years, would have to endure famine, drought, and both harsh cold in winter and roasting heat in summer. In captivity, it can have leafy green veggies and cool soaking baths in summer and warm sleeping shelters in winter. The tortoise will then mushroom in size, reaching a bulk in just a few years that would have taken fivefold the time in the wild. This outsized, corpulent, happy tortoise will very often sicken and die of disease. Like a person fed a fatty, sugary diet and deprived of exercise, our tortoise doesn't thrive on abundance. The tortoise's biology is such that it may give no outward sign of poor health as it balloons in size, but its physiology is not built to handle rich, sugary fruit and vegetables, nor constant moderate temperatures, nor ever-available water. It's hard to document such health risks because the tortoise will live ten or twenty years or more, and autopsies are rarely performed except in zoos. But a fat, overfed tortoise in captivity is much like a modern urban human being; living in an environment of such overabundance is nearly as toxic to health as the harsh vicissitudes of Nature.

But longevity and lifelong fertility cannot come close to replenishing tortoise numbers in the face of intense, ongoing depredations by people. Congdon found that although adult Blanding's turtle mortality was low, it was essential for mortality among juveniles also to be low to avoid population decline. Unfortunately, outside the protected waters of the nature reserve where his study was carried out, both adults and juveniles are at risk. Although people in affluent western societies don't as a rule eat turtles (although this is subject to fashion, as we will see), cars kill innumerable turtles each summer as they cross the road in search of nesting sites. Ponds and wetlands are drained for land development or polluted beyond habitability. People catch turtles as pets or release tropical species into the wild that harbor exotic pathogens which infect natural populations. Although many species could withstand any one of these threats, none can survive in the face of all of them.

Nevertheless, it is clear that despite their slow growth and reproduction, tortoises have an amazing capacity to rebound from disasters. Intense hunting by whaling ship crews reduced the giant tortoises on the atoll of Aldabra to near extinction in the nineteenth century. The tortoises were quickly decimated as the island was discovered and exploited for its abundant supply of tortoise meat on the hoof. Taken aboard, the tortoises, some of them upwards of four hundred pounds, could be kept alive for months and butchered when fresh meat was needed.

Aldabra was mercifully far from any trade or shipping routes, had no human settlement, and was generally such a backwater that it was left entirely alone until about the 1800s. Eventually, however, ships found their way there, and although we have no idea how many tortoises those first sailors discovered, it was a large enough number to bring them back time and again. The Aldabra tortoise population plummeted as had their neighbors on other islands, until by 1900 fewer than a thousand remained. But at that point fate stepped in. Because Aldabra had been the most recent tortoise island exploited, conservationists successfully intervened just in time to save them.

At that point the natural fecundity of these giants, combined with the ability of adult females to reproduce for many decades even when no younger generation remains, took over. With protected status, the atoll's tortoise population increased exponentially in less than a century. Today, estimates place the number of tortoises living on Aldabra at somewhere between 100,000 and 125,000. A 1970s estimate of their sheer weight on the island placed their biomass at conservatively 15 million pounds of tortoise living on a mere fifty square miles of island. They exist at the highest density of any large herbivorous animal on Earth. This rapid population growth belies the live long, live slow strategy, but it may tell us something about the ability of tortoises and turtles to respond to severe population crashes. An abundance of forage available for a small number of tortoises may have helped the remaining few to surge in numbers.

Aesop's Fable

Aesop was wrong about tortoises. Slow and steady are not nearly enough to win the tortoise's race against extinction. From a conservation standpoint, you might divide the animal world into two groups: those that can withstand human hunting or collecting for the pet trade and those that cannot. Those animals that can survive well in the face of harvesting are those that are prolific—they produce babies often enough and in enough quantity that they can sustain population levels despite extraordinary challenges. They also possess the ability to rebound from severe population cuts, so that intense hunting or harvesting would have a temporary impact but not a result that would be fatal for the species. The problem is, not very many tortoise species are built to rapidly rebound from near-extirpation. Few of them lay large egg clutches—some of the most vulnerable varieties lay only one egg at a time. As we've seen, they tend to lead long lives if they survive the perilous early months and years. If hunters remove many of the adults in a population, the remaining survivors may hang on until the end of their lives, but never repopulate their species.

First and foremost, when you're not very mobile and you find yourself quite alone, finding a mate can become a major obstacle to breeding. In the best of times, it's not easy for a tortoise to find a mate. Scientists are not quite sure whether it's done by the billiard-ball effect—randomly bumping into an attractive member of the opposite sex—or if some olfactory or other factor is at work. Studies of box turtles in the eastern United States have shown that a male and female can be only a few yards apart and not detect one another, which spells doom for populations reduced too far by human collecting. In our study of Asian forest tortoises in Thailand, we've seen one tortoise following the trail of another, many yards behind, suggesting some sort of scent trail may exist as a location cue.

When many adults are removed from the population, hatchlings and juveniles will be left behind. These spend their early lives hiding, too munchable by a wide range of predators to be cavalier about exposing themselves. For a decade or so, an area could be devoid of adults but still have an invisible population of young tortoises, which would eventually replace the lost adults. The juveniles' habit of disappearing is in a sense a safety net for a tortoise population, since seemingly extinct populations can suddenly re-appear, as though by divine intervention. But whether enough juveniles have been left to truly restock a whole breeding population is an open question.

The problem with relying on baby tortoises to replenish a population or even a species that has been hard hit by people is that their survival odds are low. That wonderful shell is not thick or solid enough to prevent a bird or small mammal from cracking it, or even a snake or large lizard from swallowing it whole. Once a decent size is reached, survival odds blossom. If ever there was an animal that could benefit from a head start early in life, it is the tortoise.

A Head Start in Life

Long lifespan and the extended reproductive issues tied to it are critical to solving the tortoise extinction crisis. Conservationists are trying to find ways to circumvent those issues in the natural design of tortoises that work against them in the modern world. These are high mortality of hatchlings, a lack of parental care by the mother who laid the eggs, a long juvenile period of slow growth, and a late start to reproduction. Their life history is designed to compensate for high mortality early in life by high survival rates and repeated reproduction late in life. If you can get the animals past the hatchling stage of maximum risk, you've dramatically boosted the population and the likelihood that tortoises will continue to have a healthy population growth even after being depleted by human activity. They just need a head start to get through their most vulnerable period, and that is exactly what many conservationists attempt to give them.

We know from field studies that head starting does work. The largest hatchlings tend to have higher survival rates in Nature even without human help. But conservation projects typically raise baby tortoises to a year or more in age or to a certain critical size, above which they are relatively immune to smaller predators or introduced pest animals. They then release them into a proper habitat after acclimating them to the harsh conditions of their natural environment.

We also know, however, that giving turtles and tortoises a head start is not a panacea. Studies of sea turtles have shown that head starting cannot prevent population crashes because it is survival in later life stages that is still critical. If you keep collecting adults that would otherwise lay eggs every year, no reduction in infant mortality can compensate. Having many chances to replace yourself in a population by laying eggs many, many years in a row may be more important than having your offspring survive in any one year. So head starting may work in certain cases and places but not in others. The current thinking is that when a species or population has been reduced to critically low levels, anything that may increase the number of individuals that may someday reproduce is worth trying.

The success of such conservation efforts is measured by the rate at which babies survive in captivity compared to the rate at which they would have survived in the wild without human care. If you only marginally increase the number of eggs laid that actually end up as adult turtles, the program may not be worth the time and money it takes to run it. If you can increase the survival rate using head starting to, say, five times the natural survival rate, you have found a key to success for the species.

At least a few studies have tried to quantify the success of head starting. Susan O'Brien and colleagues at the University of Cambridge studied its effects on one of the world's most critically endangered tortoises. The ploughshare tortoise of Madagascar has been reduced by habitat loss and by collecting for food and the pet trade to just a few hundred animals in the wild. They are aptly named for the enormously elongated front underscutes of the male's shell, the gulars, which protrude like the front tines of an old-fashioned plow. The males use these appendages as lethal weapons, getting their gulars under a rival male during courtship combat and flipping him over to die in the hot sun. They naturally inhabit dry, spiny brush areas on the western coast of the country. Today all are in confined, protected reserves to prevent poachers from pinching the very last of them. O'Brien collected and marked more than one hundred fifty baby ploughshares in the wild, all of them less than one year old at the time of capture. They then followed these babies through early life, comparing the effect of body size on future survival rates. Larger babies were more likely to survive their first year than smaller babies of the same age. Babies that reached larger size in the first year of life did so because they were larger right out of the egg, not because of any difference in growth rate during that crucial first year of life.

Why is being big better? When you're a hatchling tortoise, you may be better able to climb over obstacles or burrow under them. Your shell may be a tiny bit harder and protect you more effectively against water loss during dry periods and against bites from small predators. Among tortoises and turtles, size is as much a reflection of the environment around them as it is of age. If you have abundant resources, you grow much more quickly and reach sexual maturity earlier. This places an additional premium on rapid early growth.

Even at the earliest stage of life, size matters. In any given species, bigger tortoises lay bigger eggs in larger clutches. Combine this fact with O'Brien and colleagues' observation that larger babies are more likely to survive the first year of life and you must conclude, as they did, that the very largest females in a population contribute disproportionately to the next generation. From an individual standpoint, this is evolutionarily interesting; it suggests that just as male tortoises need to grow large enough to successfully battle other males for mates, a female's evolutionary success is strongly influenced by her ability to grow up quickly, and to a large size. The same is true for many types of egg-laying animals, such as frogs and fish, in which a female's body size translates into her ability to breed prolifically.

The most famous cases of "head starting" are among Galápagos tortoises. As we shall see later on, these giants are impervious to everything except human hands and the occasional volcanic eruption when adults, but as youngsters they are prey to all manner of predators, especially those critters intentionally or unwittingly introduced by the early human settlers of the islands. Rats eat their eggs, and feral cats, dogs, and pigs eat their young. By incubating eggs in captivity, the tiny giants get a safe and secure start. They are released on the islands of their ancestry when they reach a size of about eight to ten pounds, beyond the gape of a rat's or cat's incisors. In this way thousands of releases have been made, with many thousands more planned.

Projects such as these have a valuable role, so long as we stay aware that it is always far better to save large, breeding adults than rely on juveniles to repopulate an area. But when those adults are gone or nearly so, protecting and releasing young could in theory be practiced en masse. Think of all those species for which big captive populations exist. All it takes is concerned citizens to donate some of their baby tortoises for releases, and large-scale reintroduction could take place. We might envision a future in which the breeding of pet tortoises would lead to available populations for reintroduction to the wild. Such projects are currently being conceived for desert tortoises in the American Southwest. There are many barriers to the success of such programs, such as the risk of introducing diseases or of altering the genetic structure of wild populations. And such conservation efforts only work when the species in question is prolific. Those tortoises that lay just an egg or two at a time would not be able to provide enough captive-raised progeny to effectively repopulate their species.

If a baby tortoise clears all the hurdles between conception and the first few years of life, it enjoys the prospect of extreme longevity. It is commonly thought that such ancient creatures must take near eternity to reach their full size, but in fact tortoises are capable of amazing growth rates. They possess the ability to turn a huge percentage of the calories they eat into body mass. In the wild, they grow in accordance with the environment; putting on length and weight during rainy, lush periods, and growing little or not at all during droughts and dry seasons. A baby Galápagos tortoise that is three inches long and weighs a few ounces at hatching will, in twenty years, put on hundreds of pounds and a couple of feet of length on a diet of mostly grass, scrounged and grazed during most waking hours. In captivity, with little to do but eat, that same growth—from a mini-tortoise that can nestle in your palm to a behemoth a teenager could ride in ten or twelve years—can be achieved in a decade or less. But this rapid growth is not necessarily healthy for the tortoise. They evolved for slow growth, and there is much evidence that the low-fiber, high-sugar, and high-protein diet of market vegetables presented en masse may have a harmful effect. An animal designed for longevity may compromise that very thing by putting on weight and growing so rapidly.

This is the tragic irony of the tortoise. Everything about it is built for the duration, but contact with humans tends to have swift consequences for which the creature is utterly unprepared. Whether we kill them with kindness or malevolence, we destroy them in greater numbers every year. In the following chapters we will look at some of the most sinister threats facing these peaceful relatives of ours and consider whether their fate is to disappear from Earth like so many primitive creatures before them, or to be preserved in some form by their main nemesis.

3

NO RESPECT FOR THE ANCIENT LANDS

Wal-Mart is coming to Lake Park. The chain will build a cavernous superstore in central Florida. The land has been acquired, the permits have been pulled, the construction crews stand ready. One last little annoyance: there are gopher tortoise burrows on the land. Wal-Mart has to deal with land use issues far more difficult than this in their quest to offer America cheap retail products. But the tortoises, unlike many small businesses and other private interests, have passionate, vocal advocates, and therefore state officials who act on the animals' behalf. Wal-Mart, however, knows what a bureaucratic hassle it can be to deal with delays, potential work stoppage, and design plan changes that result from finding endangered animals in the path of your backhoes. So they pay $11,409 to the State of Florida for the right to bulldoze the tortoises into their burrows, entombing them. An entombed tortoise will struggle in the packed soil under which he has been buried for weeks or months before succumbing.

The death toll in Lake Park may have been as low as five tortoises. This is dwarfed by the approximately 75,000 gopher tortoises buried in such development projects over the past decade and a half in Florida alone. The same upland soils in the southeastern United States that make perfect gopher tortoise habitat are unfortunately also prized for building. Every time a housing subdivision or shopping mall is built, developers have been able to sign away their obligation to protect endangered animals by flashing some cash, and states have been willing to let them get away with it. The cash is used by the state to purchase and maintain land elsewhere that can in theory protect other tortoise populations in perpetuity. But is the cost of the land worth the sacrifice of burying alive those tortoise populations that stand in the way of never-ending new development?

There are alternatives, but most are ineffective. The developer can opt to fund a translocation of the tortoises living on his building site. Unfortunately, such efforts are more effective at enhancing the public reputation of the developer than they are at saving the tortoises. The local media will document the noble effort to rescue dozens of tortoises from the path of the bulldozers, but all evidence suggests that the vast majority of translocated tortoises die unpleasant deaths not long after being unceremoniously dumped into "suitable proxy habitat." They're not rocks, despite appearances, and tossing them into what looks to us like tortoise-friendly land doesn't mean they will adapt and thrive. Disconnected from their ancestral home, they tend to wander aimlessly in their new, completely unknown surroundings, and most end up as roadkill.

Eco-Blowback

Habitat loss is the elephant in the room when it comes to wildlife conservation. People destroy habitat when they turn forests and fields into farms, villages, lumber, pastures, and roads, and when they turn ponds into livestock waterholes or waste-runoff basins. It is a consequence of development, which is itself a consequence of population growth. In many areas of the developing world, population growth is increasing dramatically, so cutting birth rates will affect the rate of growth but have little impact on near-term overall population increase. There is no question about it; in fifty or a hundred years, Earth will have far less natural space, and what remains will be more degraded by people. We don't know when and how this will end, but a reasonable extrapolation would be that in the twenty-second century, the planet's population will be at least twice what it is today, perhaps 15 billion. Remaining fragments of forest and natural habitat will gradually become a crazy quilt across the Earth. The industrial countries have already established the vast majority of their national parks, state parks, and other protected areas, and although the level of protection varies, their inhabitants are relatively safe from human development. In the developing world, where most tortoises and turtles live, areas protected by signposts are many, but enforcement is often nonexistent or ignored.

The risk to Earth's wildlife seems obvious, but what is less discussed is the hidden impact of habitat destruction on people—the unexpected blowback, as it were. Many of the world's tropical forests stand on soil that is not capable of supporting large-scale agriculture. In the absence of trees, the earth is rapidly eroded by torrential rains, washed bare of its essential nutrients. Loss of habitat for wildlife and loss of arable land combined with exponential population growth in the developing world is a recipe for humanitarian disaster.

Without viable alternatives, the world's poor will continue to depend on foreign aid from the wealthiest countries for the foreseeable future. And the warming effects of greenhouse gas emissions have shown us that large-scale human alterations of the environment can lead to results worldwide that are difficult to change.

For tortoises, this worldwide prognosis is bleak. At the current rates at which tortoises are consumed, made into pets, or lost through habitat destruction, their fate a century from now can be predicted, country by country.

In regard to conservation, *China* is thus far the black hole of Asia, exerting a gravitational force that sucks in smuggled wildlife from the entire world, destined for the stomachs of over a billion people. The taste for all things chelonian and the crush of overpopulation will render virtually all wild turtles and tortoises extinct, and natural habitat will remain only in a few well-protected areas. A precious few animals will be kept like crown jewels in locked enclosures, to be trotted out for visiting biologists (and perhaps culinary experts). Many millions of turtles will be seen in markets, all of them farm-raised. Likewise, *Vietnam* will soon have no turtle or tortoise fauna to speak of, and with a burgeoning population problem, the same applies to most other animal species there.

Thailand has the best chance to be the major repository of biodiversity in Southeast Asia, thanks to its increasing environmental awareness and substantial number and size of protected areas today. Increasing numbers of middle-class Thais go to the forest to camp, take photographs, and paint watercolors of sunsets rather than hunt.

Indonesia presents a heartbreaking scenario; many of its spectacular natural areas are nominally protected, but the government either tacitly or actively (and often illegally) permits logging in order to line its pockets. Sometimes the forests are cut and then replanted with vast oil palm plantations, giving them a veneer of green that is in fact useless to wildlife. The biodiversity of Indonesia is enormous, the wildlife unique, but population growth and ongoing rampant deforestation will likely mean that a century from now, only the tiniest of fragments will exist.

If China is the black hole of Asia, then *Myanmar* (Burma) must be the abyss. On the one hand, the Burmese landscape is still largely covered in its original forest, and because of its location between the Indian subcontinent and the rain forests of Southeast Asia, it has the best of both worlds in terms of Asian biodiversity. On the other hand, it has one of the few governments in the world that is unwilling to be a party to the treaties regulating the international trade in wildlife. The government is hell-bent on cutting down every tree in the nation; they're just so destitute and inefficient that it hasn't happened yet. Only a change of government accompanied by an infusion of foreign aid could stave off the destruction.

Madagascar is already so ecologically impoverished that you can drive hours and hours and see no animal life at all save chickens and cows. Crushing poverty, lack of infrastructure, and remote location limit the success of ecotourism. Only those tortoise populations that coexist with lemurs, which are large and splashy enough that ecotourism projects and nature reserves can be built around them, are likely to survive. The habitat available to the little flat-tailed spider tortoise, or *kapidolo,* has shrunk by nearly half since the 1960s. The rate of deforestation is, as nearly everywhere else, increasing rather than decreasing.

A conservation island in a continent of trouble, *South Africa* is one of the very few African nations that has the resources and wherewithal to effectively manage its wildlife. As long as the country remains a model of democracy and stability in comparison to its African neighbors, the many parks and reserves, including a few set up specifically for tortoises, should be saved in perpetuity.

Wherever large, glamorous mammals that have tourism value live, any small creatures living in the same habitat will gain incidental protection. So tortoises in *Kenya* and *Tanzania* will benefit in nature reserves, although outside protected areas and all across the rest of the African continent, even the largest species are threatened. The little pancake tortoise may be an unlucky exception; even inside nature reserves poachers continue to break into their rocky lairs to steal them for the pet trade. Their reproductive potential is low, and their long-term future is therefore quite bleak.

In *Europe* the myriad species and races of Mediterranean tortoises of the genus *Testudo* will suffer until local trade is greatly curtailed. The continent with arguably the greatest awareness of the welfare of tortoises will see many of the local varieties of its chelonians disappear.

Perhaps ironically, the future of tortoises is most uncertain in the *Galápagos Islands*. On the one hand, it's a famed protected area with a booming tourism industry built partly on the sturdy backs of giant tortoises. On the other hand, ecological problems and conflicts abound; as recently as 2000 the National Park headquarters was taken over and ransacked by angry fishermen protesting conservation measures in the islands. With each small island holding one or a few distinct races of tortoises, each critically endangered, the long-term future will always be precarious. So long as Ecuador remains mired in poverty, migrants will come from the mainland seeking employment, and the pressure on the tiny ecosystems of the archipelago will be intense.

The two largest tortoises on the mainland of *South America,* the red-footed and yellow-footed, are not critically endangered at present, though they are still widely collected for food and for export as pets. As the vast forests of the Amazon Basin shrink dramatically over the coming century, tortoise depredations will likely outpace habitat loss, but these species have at least a moderately long-term survival prospect in many areas.

In both the southeastern and southwestern *United States,* our native tortoises will be preserved on protected lands, but they will cease to exist everywhere else. The battle between developers and tortoises almost always ends in favor of the real-estate deal, but clever negotiating, land parcel by land parcel, will leave viable populations of native American tortoises for the long-term future. But this will be possible only if tracts of land that are large enough and far from freeways are preserved, and if current concerns about the vulnerability of tortoises to epidemics are allayed.

The prognosis for tortoises may sound bleak, but two factors offer some hope. First, these extrapolations are based on current rates of habitat destruction, current levels of affluence versus poverty, and the current political stability or effective policy-making by governments. Any of these factors could change. Governments like the one in Myanmar may fail, and a democratically elected government open to global conservation initiatives may come to power. But governments sympathetic to conservation may also fall into chaos—see Zimbabwe. Sometimes the only silver lining for conservationists in a third-world dictatorship is that the rules are actually enforced; when a dictator says poaching will be punishable by death, the decree carries weight. In democracies, battles rage for majority consensus on how best to manage and conserve natural resources, including rare wildlife.

Second, all habitats are not the same. In some cases, a species may thrive on a small area of land, which if it lies in a well-protected zone can represent safety in posterity. We will see later that the geometric tortoise in South Africa is one such case. The radiated tortoise, that big beauty queen of Testudinidae, is often thought of as teetering on the edge of extinction. Its official conservation status reflects this. But several surveys and censuses have shown that in the proper habitat and with strict antipoaching enforcement, they live at stunningly high densities. Walk through a scrubby forest in southernmost Madagascar during the rainy season and you will rarely be out of sight of a magnificent radiated tortoise plodding along. This is presumably their natural density; under the right circumstances they are as abundant as ants. But travel a few miles down the road to a less suitable habitat and the density drops. Go another several miles to areas where people have been collecting the tortoises for food and pets and the density drops still further. Although populations of radiated tortoises have been measured several times in recent years, the results have varied wildly from one study to another: one indicated as many as 10 million tortoises remaining, another a small fraction of this. What we don't know is whether the immense densities seen in some sites are in any way vital for the survival of the species. If, for instance, males and females had trouble finding the right mates in anything less than optimal densities, then even a modest drop could have far-reaching effects. For those species that thrive only at high densities, even moderate habitat loss or poaching can be devastating.

Other tortoises have the opposite problem. They occur over a wide geographic area, but are nowhere common. They may have highly specialized habitat requirements, and if those particular habitats are valued by people, they are in trouble. The beautiful impressed tortoise occurs from Myanmar across Southeast Asia and southernmost China into Malaysia. But across that large area it is found only in patches of mountainous wet forest with a certain mix of bamboo and evergreens, with the right temperature range and rainfall. That adds up to a patchwork quilt, with vast areas of the tortoises' theoretical range consisting of completely unsuitable rice paddies, plains, and towns. As agriculture and settlements have cut up the forest, proper impressed tortoise habitat has become a loose chain of high-elevation islands, forever isolated from one another. The species would at least be safe in their highland realms if people would leave them alone there —the areas are typically too precipitous for decent farming anyway—but they are hunted for food by local villagers.

Unsustainable Development

It's politically incorrect to be against development in developing nations, which is supposed to improve the lives of those residents mired in a level of poverty that no western person can truly comprehend. Conservationists tend to be vocally against population growth but recognize that it's awfully difficult to do much about it. So as a salve, conservationists turn to the concept of sustainable development, in which people living near wildlife areas would increase their standard of living without thoroughly trashing the land and its animal inhabitants. It should lead to people having more money and a higher standard of living. Their increased economic stability would make them less likely to opt for short-term gains like cutting down all their trees in favor of long-term investment in the land. Local people would be given an economic incentive to respect the forest, and as their economic standing increases, they would be more likely to use the forest in a sustainable, lower impact way. Ecotourism, for instance, might replace hunting as a local industry, and local people might be employed as guides instead of as poachers. The forest benefits, and ultimately people benefit. "Conservation without illusion," according to Jonathan Adams and Thomas McShane, should appreciate that people have co-existed with wildlife for millennia. Trying to exclude people from wildlife habitat is, they argue, an ultimately counterproductive attempt to turn back the clock to a time that never really existed anyway.

The goal of sustainable development has spawned a global industry of projects and funding organizations. "Sustainable development" is a buzzword to trounce all buzzwords on the floors of every governmental developmental agency. But to some conservationists, sustainable development is an oxymoron that portends little but disappointment. Giving rural people in the developing world a role in managing their natural heritage sounds politically correct, but does it work? The answer is both yes and no. Plenty of studies have shown that, despite what we'd prefer to believe, people tend to exploit resources in their local environment for maximum short-term gain, not long-term management. This applies to cultures from Native Americans to Amazonian Indians to, of course, European settlers on all continents. For people living along the margins of a natural area that harbors wild animals, a higher standard of living may encourage them to turn to other sources of protein, such as beef from the market or from their farm. Or it may simply enable them to use their newfound cash to buy guns and bullets, which are far more effective at procuring protein than bows and arrows. Sustainable development may be a goal and an achievement in some times and places, but as a panacea it may be chimerical, a feel-good effort with the potential for negative repercussions.

How Much Land?

Habitat loss is not an equal-opportunity killer. Some species can cope with or even thrive in changing landscapes. The number of white-tailed deer in the United States is exponentially greater now than it was in precolonial times because so much continuous forest has been cut up into small plots, leaving mile after mile of forest/field edges full of the leafy plants deer favor. A few bird species—robins, for example—that prefer open parks to dense forest have also flourished. These "weed species" are habitat generalists, adapted to the border zones between two habitats or even to degraded habitats that somehow resemble ecological boundary areas. Habitat specialists meanwhile go extinct in the face of such radical landscape alterations. Most if not all tortoises are specialized to the extent that extensive environmental degradation, not to mention outright habitat destruction, will kill them off. In some cases, removal of just certain plant species—through logging, for instance—can have a severe impact on the animals in the forest even though most tree species remain. In the case of aquatic turtles, runoff of chemicals into lakes and ponds can destroy habitat even though the pond is as scenic as ever and the surrounding lands appear unpolluted.

One depressing realization in the habitat loss problem is that it's not enough to protect natural habitats. They also must be managed, even proactively manipulated. For gopher tortoises to survive in the longleaf pine forests of the southeastern United States, a certain cycle of Nature must be kept unbroken. Longleaf pine forest is parklike; for millennia, fires started by lightning strikes have cleared undergrowth, enriched the soil, and controlled the invasion of other tree species not adapted to the cycle of fire. The tortoises rely on the open park spaces to make their burrows and find their preferred plant foods. Federally protected lands often have controlled burns to renew the cycle and restore the dynamic balance in the ecosystem. The tortoises wait out the fire deep in their burrows. But owners of private land don't like to set fires, so their stands of pine tend to become choked with brushy undergrowth that renders it undesirable as tortoise habitat. Since they're also not allowed to develop the land if it means killing tortoises, they're stuck with a lose-lose situation; tortoise populations keep dropping over time while the owners watch, unable either to help the tortoises or to exploit their own land.

A few conservationists in the longleaf regions of Alabama started thinking out of the box and came up with a solution. Landowners are allowed to invest in a "conservation bank"; a parcel of degraded longleaf habitat that is managed and restored using funds they themselves have donated. Once the forest has begun to look like tortoise habitat again, tortoises are relocated there from the owners' land, leaving the owners free to do what they choose with their property. A lose-lose situation became a win-win one. Instead of holding a grudge against the tortoises, property owners see them as an investment. And the government agencies involved come off looking clever, which is rare. This sort of fix won't work everywhere, but it highlights the need for action on lands that don't belong to the government, where the endangered species regulations bite hardest and with the worst aftertaste for property owners.

The real lesson of conservation banking is that novel solutions are required to cope with the foremost habitat loss issue in many developed countries: fragmentation. Vast tracts of forest that were once continuous are now sliced, diced, and chopped into ever-smaller fragments. The effect of this is not only the obvious direct one: reducing the overall amount of habitat for tortoises and a host of other animals. The really insidious effects of fragmentation are often invisible unless you know what to look for. In developed nations, many communities have woodlots that are the pathetic remnants of once-mighty woodlands, now reduced to postage-stamp size and maintained lovingly by local communities. But when you reduce the size of a forest plot, then by definition you also increase the amount of forest edge. The forest edge is where a lot of animals make their living, hence the explosion of deer, squirrels, and a few open-country bird species. Meanwhile, the deep forest dwellers disappear. And the more edge you have, the more penetrable your forest becomes. Studies have shown that when woodland plots are under a certain distance from edge to deep interior, the entire plot is at risk from marauding dogs, house cats, and the like. The family tabby can annihilate the entire nesting bird population of a small woodlot in no time. Imagine the impact of dogs on turtles and tortoises, not to mention the greatly increased odds that the entire tortoise population will wander outside the forest to be smashed by cars or taken home by acquisitive pedestrians.

Studies have shown that, even in areas far from human settlements, forest fragment size directly affects the diversity of animals living in that fragment. In a landmark experiment, Thomas Lovejoy and his colleagues cut forest blocks of varying sizes between one and one hundred hectares (about 250 acres) in a rain-forest area of Brazil. Over the following fifteen years they monitored changes in plant and animal life in those plots. The finding was startling. In even the 100 hectare fragment, half of the animal species that had been there when the plot was created were gone within fifteen years. This suggests that it is too late for conservation measures to save biodiversity in any small forest fragment, since the politics, funding, and scientific research required before effective measures could be put in place would just take too long.

Forest fragments are very much like islands, except they are isolated by fields, farms, and towns instead of by water. And they lose species in the same way that small islands do. Although a small fraction of the world's animal species live on islands, most well-documented extinctions—perhaps as high as 90 percent—have occurred on them. A tropical island in Panama, Barro Colorado, came into being when the valley of the Chagres River was flooded in 1914 to create the Panama Canal. In its earlier life, the new island had been the peak of a hill. In the century since its creation, Barro Colorado Island has been meticulously studied and is probably the best-known tropical forest, inch for inch, in the world. Although it is closely protected, there are no longer any large mammals—jaguar, tapir, deer—on the island, and over the decades, the diversity of small animals has slowly declined too.

How forest fragmentation affects tortoises is not always clear. A study of yellow-footed tortoises—large forest-floor dwellers of the Amazon Basin—showed that forest fragmentation led, in just a decade and a half, to altered mortality rates and life-history trajectories. In a 350 hectare forest fragment, there were far more young tortoises relative to adults than one would find in nearby larger and continuous forest blocks. There are a couple of hypothetical reasons for the reduced adult population: adults are more easily found by local people, who eat them; adults may be more likely to wander out of smaller forest blocks. On a more cheerful note for tortoises, the disproportionate number of juveniles to adults may indicate decreased mortality in young tortoises because predatory animals that eat eggs and babies disappear from small forest blocks quickly, as we have seen.

Not all species respond in the same way to fragmentation. A study conducted in the coastal scrubland habitat of the gopher tortoise in Florida found that although the abundance of most vertebrate species corresponded to the size of the forest fragment, more than a third of all species did not suffer from being squeezed into smaller habitat blocks. This was because small clusters of forest blocks in proximity to one another served as decent enough surrogates for one single big block. Gopher tortoises seemed to be more sensitive to the structure of the forest and the exact composition of plant and tree species than to tract size overall.

One way to mitigate the sinister effects of habitat fragmentation is with habitat corridors. The general idea is that if you offer wild animals strips of their natural habitat as connective tissue across deforested land, or across natural forest that has been converted to timber plantations, the animals will make use of them to live in more than one forest block at the same time. You've likely seen habitat corridors at work; many freeways have tunnels under them to allow deer, bear, and other wildlife to get from one place to another without being hit by cars rushing by. So what if you maintain a narrow path of forest through areas that are otherwise unusable by tortoises? It appears that, at least as a last resort, this can work. Land owners can work together to manage their adjacent properties for wildlife if they agree to preserve, or to replant, corridors that turn their dispersed fragments of forest—perhaps land that is unusable for other purposes—into a web of linked tracts of natural habitat.

Unfortunately, more often the corridor effect works in the opposite direction; the corridor is not created as a safe haven but as a linear clearing for power utility towers or a spot for underground pipe tunnels. The resulting hash marks are known to disrupt natural movement patterns of smaller animals, effectively turning an otherwise nicely continuous forest block into a series of fragments. Although the building and installation of power lines no doubt kill many animals outright, including tortoises, the indirect effects of fragmentation are less well known. Yet the most ubiquitous and deadly of all corridors that intersect tortoise habitats is roads. Studies of desert tortoises in California have shown that otherwise healthy tortoise populations are depressed for a quarter-mile on either side of a roadway stretching across the desert, presumably due to road kills. Even our best desert wildlife reserves are cut by plenty of roads for public access.

Road Kill

Road kill is the butt of many jokes, a synonym for an easily defeated opponent in football or political elections. But road kill, the real thing, is also a dramatically lethal cause of death for wildlife, whether as large as a grizzly, giraffe, or elephant or as unexpected as a bald eagle. It is therefore not surprising that tortoises often end up under the wheels of a vehicle. It's a global problem, and the relatively few studies that have been conducted show that it can be the leading cause of death in areas where you wouldn't expect it. You might not be shocked if I say that an estimated million vertebrates per day are killed on the 4 million miles of American roads. But what if I say that in the great savannah parks of East Africa, more giraffes, baboons, and elephants are killed by cars than by lions? That half of the remaining few Florida panthers (also known as mountain lions) have been killed by cars over the past twenty years? That some endangered species have disappeared from their natural ranges and cannot be reintroduced, not because of hunting or habitat loss, but because of the risk of road kill?

A high-speed roadway through wildlife habitat is a guillotine that randomly slams down on animals unlucky enough to pass through its maw at the wrong moment. But such deaths are more than random for some animals; roads are often attractive nuisances. For a tortoise, the heat-retaining asphalt may be the best source of warmth in the morning hours, when the surrounding desert is still chilly. For a female looking for the right place to lay eggs, or a male looking for a female, a roadway is a lethal obstacle that must be crossed. And a road-killed tortoise is bait for scavenging mammals or birds of prey that settle onto a crushed tortoise on the roadway. These may end up as road kill themselves. Driving across southern Australia, I saw hundreds of kangaroo road kills. And on top of many of them, there were big feral domestic cats—tabbies gone wild—tearing into the kangaroo carcasses. More meat on the landscape means ready food for the cats. But a population explosion of domestic cats could not be worse for local wildlife, tabbies being ferocious predators on a host of small, and often endangered, wild animals.

There are easy ways to lessen the impact of road kill, but these are often prohibitively costly. A study in Arizona showed that simply placing fencing along stretches of desert that border major roads reduced road-kill tortoise deaths by 75 percent. Such fences have to be funded and put up by someone and can cost over a million dollars, but if done in the right areas, they have the ability to dramatically mitigate the lethality of roads for wildlife.

Reintroduction Fantasies

Once upon a time in the mid-twentieth century, it was a well-established tradition for suburban southern California families on vacation to pick up desert tortoises and bring them home to become backyard pets. Perhaps you would be camping with the kids in the Mojave Desert, the spring wildflowers bursting like tiny fountains of color everywhere. As you headed home, you'd spy a big old tortoise ambling across the blacktop. Next thing you knew, the tortoise had been "rescued" at the behest of your kids and was a passenger traveling at high speed toward a new home in which lawn grass and ample heaps of veggies substituted for the animal's natural diet of scrubby weeds, and lawn sprinklers took the place of cherished winter rains. Life was good, and easy. Tortoises were plentiful, both in the desert and in their translocated homes in suburbia.

Conservationists were appalled at the lightning speed at which this slow-moving animal was losing its habitat to development, its lives to speeding cars, and its home to pet-hungry families. Biologists worked hard in the 1960s and 1970s to set aside desert tortoise habitat in the southwestern United States to protect dwindling tortoise numbers. Small, fragmented sanctuaries were established. Legislators did their part, and legal protections were enacted. In 1973 the federal government listed the desert tortoise (*Gopherus agassizii*) as "threatened" under the Endangered Species Act. With considerable effort, more populations representing about a third of all desert tortoises were given protection in 1990. Large tracts of desert land were locked away in perpetuity for the preservation of the desert tortoise.

The legal protection for the tortoises and the media attention surrounding it was clearly a good thing for the future of the species. But it may have had dramatic unintended consequences. When tortoise owners learned that the pets they had taken from the wild years earlier were now on the endangered species list, many responded in the best-intentioned way possible. They put the tortoise in the car, drove him back out to the desert, and released him with a sense of having done something good in the world.

And then the law of unintended consequences took over. Within a few years of the initial listing of desert tortoises as threatened, biologists began to find wild tortoises that were sick. They had runny noses, sunken eyes with puffy lids, and an unhealthy lack of color in their skin. The syndrome came to be known as Upper Respiratory Tract Disease, or URTD, and it closely resembled a disease that had been noticed for many years in captive desert tortoises that had been kept in contact with pet tortoises of other species. No single pathogen had ever been isolated for the captive disease—it wasn't even clear if it was bacterial or viral in nature—but it was often fatal. In Europe, where most pet tortoises hail from the Mediterranean, the disease had taken a heavy toll. What's more, the area in which the disease first appeared in wild desert tortoises—the southwestern corner of the state of Utah—was the same place where many pet tortoises were known to have been released in the preceding decade.

Research was begun into the disease and its impact on wild tortoise populations. A *Mycoplasma* bacteria—the same group of bacteria that can cause pneumonia in humans—was implicated as the main culprit, with some possible bacterial agents being involved as well. Although poor nutrition (such as happens when a wild tortoise's habitat is degraded by human use and food plants suffer) may be one cause, there is little doubt that a major source of the disease in the southwestern deserts is the pet tortoises released into the wild that carried the infection to large numbers of healthy wild tortoises. As in humans, the tortoise *Mycoplasma* (since identified as *Mycoplasma agassizii*) is difficult to treat and not as responsive to antibiotics as other genera of bacterial infection. It's a chronic illness, and even if the tortoise copes with it, it may recur in times of stress or food shortages. In tortoises, the symptoms can be dealt with and the animals restored to health with a diligent course of antibiotics, but they remain carriers of the disease for life.

The theory that released pet tortoises are the culprits in the *Mycoplasma* epidemic has its detractors, who claim that the tortoises may have co-existed with the disease for millennia in Nature. When habitats become degraded by human actions, these conservationists believe, the tortoises' immune systems are less able to suppress the health tribulations that have always been present.

There are now tortoise reserves that were established in the desert decades ago that are for the most part empty of tortoises. The massive die-off of tortoises, mainly in the Mojave Desert of southeastern California, has emptied thousands of square miles of prime, often protected tortoise habitat. Instead of densities of hundreds of tortoises per square mile, some reserves now have tortoise densities of only a few dozen per square mile.

The disease problem also affects our ability to translocate displaced tortoise populations. When a military base announces plans to expand its missile target range or its training areas, it must file an environmental impact plan that accounts for the wildlife that will suffer in its wake. Tortoises could be rounded up and moved to another area out of direct harm's way (though we saw earlier that is no panacea) but for the disease risk. You can't risk infecting a whole new population by introducing diseased tortoises. So many may have to be quarantined long term, which adds another unintended consequence.

Challenges from Every Side

Disease is hardly the only problem that desert tortoises face. Wildfires, once a rarity in the arid desert lands, have become more and more common as nonnative plants have invaded, providing ready kindling. Coyote populations have boomed and coyotes are ruthless tortoise predators, as are ravens. Ravens, always a feature of the southwestern deserts, have long been predators on young tortoises. But ravens have undergone a recent population boom as development infringes on the wild desert and garbage dumps balloon in size and number. And of course there are the ubiquitous off-road vehicles, their owners lobbying for more and more lands to be eaten up by their wheels, their sport flourishing in spite of, rather than in tune with, the natural world. In the name of high-decibel fun, they crunch tortoises and flatten their burrows, leaving little more than fumes and desolation in their wake.

A more subtle but far-reaching threat from humans is the alteration in tortoise diet that development causes. Research on tortoise diet has revealed that in low-rain years—that term is relative in an area in which an inch a year is a lot— the tortoises subsist on practically nothing. A few nibbles of withered cactus and dried weeds must suffice for months on end. Then, when the occasional wet year arrives, the animals become highly selective feeders. They choose only low-potassium foods. Potassium is a potentially toxic substance for tortoises. Normally, they excrete it to the extent possible to avoid a lethal buildup. They consistently try to eat foods that are protein rich and potassium poor. Protein can be used to assist in ridding the body of potassium, but then it's not available as a badly needed nutrient. So when the rains come and food is abundant, the tortoises must bulk up on high-protein, low-potassium foods to cope with the food shortage sure to come. As introduced weed species invade the desert, they often displace low-potassium native plants, and the tortoises have a harder and harder time finding anything palatable to eat.

The odds are that desert tortoise numbers will continue to decline, and that fifty or a hundred years from now their situation will be much more dire. This may seem surprising for an animal that lives in largely uninhabited, fairly useless land in a nation wealthy enough to set land aside and carefully manage it. Native American tortoises are a good example of what happens to wildlife in the face of habitat loss, if only because habitat loss in North America gives us a glimpse of the coming apocalyptic habitat loss in the rest of the world.

No Single Solution

The desert tortoise problem is the problem of wildlife everywhere. Here, in a wealthy country full of eager students who want to learn more about the animals, and with conservation biologists devoting their careers to studying and protecting them, we see that the threats to the tortoises are multilayered and often hard to tease apart. So consider the predicament in which tortoises elsewhere find themselves. Those who might wish to protect tortoises elsewhere often don't even know the full extent of their geographic distribution or their exact status within those areas. Perhaps a study or two has been done—just basic field biology. We have no detailed long-term information on which to build a conservation strategy. So things are bad and getting worse for most of the world's tortoises, for whom loss of habitat is the major source of destruction.

Habitat issues are not the same for all endangered species. For some, the habitat is well known and protected, just pathetically small or degraded by human activity. For others, the habitat remaining may be impressively large, but may lack protected sanctuaries or exist in nations that are not interested in implementing conservation action. The epitome of the first case would be the geometric tortoise, or suurpootjie (*Psammobates geometricus*). This beautiful little creature with a starburst shell pattern occupies one of the most restricted habitats of any vertebrate animal on Earth. Only a percent or two of the original scrubby habitat of the tortoises remains. The conversion of wild lands to farms and vineyards in the Western Cape Province of the Republic of South Africa, abetted by collection of the tortoises and their eggs for food by local people and illegal collection for the pet trade, reduced the population to a few thousand or so scattered in three perilously tiny fragments of land, each of which is isolated from the other. The largest of several small nature reserves established to protect the tortoises is only a few thousand acres and holds perhaps 2,500 tortoises. Most of the rest cling to existence by the handful on privately owned farms, wineries, and estates.

In one sense, the geometric tortoise is an extreme example, with so little land available and so few survivors of the species remaining. But its savior may be that its home is South Africa, the most affluent and progressive nation on the African continent and a place where conservation strategies can actually be implemented, rather than only wished for. Imagine a place where an equally critically endangered tortoise might live, except without the benefits of infrastructure, money, and concerned citizens with the wherewithal to act. Such cases outnumber the case of the geometric tortoise. The ploughshare tortoise, or angonoka, is another of the tortoises on death row, as it were: at last estimate a few hundred remain in the wild, all in disconnected forest fragments over a small area of land.

This beleaguered tortoise has a history filled with exploitation by people. They've been collected for food, stored on boats headed along the coast to be eaten when needed. Their habitat has been subjected to burning by pastoralists who hope to create new plant growth for their cattle and don't mind roasting some of the world's rarest tortoises in the process. And pet owners view the angonoka as the ultimate forbidden fruit; you know you shouldn't own one, but that makes the temptation to get one at whatever expense all the more intense.

Angonokas live in the scrubby forests of the western coast of Madagascar, not that many hundreds of kilometers from South Africa but a world away in terms of resources available to protect them. There are no middle- or upper-class landowners who can be recruited to spend some of their revenue from wineries or bed-and-breakfasts to conserve the tortoises. The wilds of Madagascar are difficult to police for poachers, and a single angonoka may sell for over twenty thousand dollars on the global black market. Those taken from Madagascar may be in the smuggling hubs in Bangkok a few weeks later. The only limits on the illegal take are the ingenuity and persistence of the thieves.

As recently as the early 1990s only a few hundred angonokas remained, and a captive breeding program was set up in Madagascar, stocked with confiscated pet angonokas. The program has been quite successful in breeding hundreds of baby angonokas, although unfortunately not as successful at preventing their theft by poachers from the breeding center. But through added security, the captive population has greatly increased, with periodic releases back to the wild now possible. Only one truly wild population exists, in a place with difficult access and so perhaps protected by its isolation. So the remoteness of the angonoka's habitat is both its boon and bane. It keeps poachers away, but for the determined thief, where there's a will there's a way, and law enforcement is thin on the ground.

The survival issues of the angonoka demonstrate that the conservation problems in desperately poor nations are different from those in wealthy ones. In the best of times and places, extinction threats due to habitat loss can be intractable. Nearly all of the original forests that once blanketed the United States have been cut down. What has been allowed to grow back is often different, and of lower quality to wildlife, than what was there before. We tend to forget that by cutting down forests we have already driven species into extinction: the ivory-billed woodpecker, the Carolina parakeet, the passenger pigeon (the pigeon was hunted intensely too, but loss of natural forest led to the crop-raiding that brought them into contact with hunters). The United States has the largest landmass locked up in national parks and sanctuaries of any similar-sized landmass on Earth. In the less developed countries, where human population densities are often much higher and where land is still divided in postage-stamp plots for small-scale family agriculture, the luxury of setting aside large tracts to preserve ecosystems does not exist. And in the least-developed countries, the balance of remaining pristine wildlife areas and human development is only tipped temporarily in favor of the natural landscape because the tools of development—bulldozers, bank loans, and government subsidies for developers—do not yet exist. The government of

Myanmar is determined to exploit the nation's vast timber re-serves; it's been limited so far by the need to use elephants in remote areas in which machinery is either not available or not up to the task.

Those who think conservationists will do anything to lock up natural landscapes at the expense of local people and de-veloping economies tend to ignore the economists them-selves. Even development-minded economic analyses show that protecting natural resources is a necessity for both poor and wealthy societies. Until we grasp the idea that forests tend to be more valuable than open plains subject to soil de-pletion and flooding, we're not going to make any progress in either saving wildlife or promoting development.

The cycle of habitat destruction and economic loss is a sin-ister one. The Indonesian government has for years cut down its rain forests, sometimes quietly sanctioning illegal logging in national reserves, in order to finance its foreign debt and line the very deep pockets of government officials. The Sa-haran areas of northern Africa advance every year, bringing human misery as arable land changes to desert through tree cutting and overgrazing. The same has been going on for cen-turies in the desert regions of India and the steppes of China. When I was living in rural Bangladesh in the 1980s, the an-nual floods reached historic proportions, and much of the na-tion was underwater. This isn't unusual in Bangladesh, the epitome of environmental blowback. But that year the floods happened mostly under clear blue skies. The rivers over-flowed all over the country because they held so much silt from horrific erosion in the Himalayas that tower above the north of the country. Those mountains used to be forested, but a combination of wholesale British colonial and Nepalese logging has turned them into a vast bare mountain front where nearly all the rain sluices down into the great rivers of the Gangetic plain, carrying vast quantities of soil with it. The rivers can't hold the annual rains any longer, and poor, flat Bangladesh frequently pays an awful price.

Of course it's a lot easier to see that environmental destruction causes human hardship than it is to persuade poor nations to restrain themselves from exploiting the bounty of Nature sitting at their doorstep. It's almost as hard to persuade the people of wealthy nations to restrain themselves from exercising their libertarian rights to make whatever use of their land that they choose. So habitat loss has been, and likely always will be, a death knell for the voiceless creatures of Earth.

4

EATING TORTOISES

The men have been walking for hours, over rugged hills and through thickets and swampy leech-filled valleys. Bamboo grows thickly on the hillsides, and it's slow going as they push through the prison of upright canes. Rusted rifles are slung on their backs. They walk through a forest that was unbroken in all directions in their childhood, but which now exists only in patches here and there. Ahead of them is the largest tract of forest in the area, much of which is protected as a wildlife sanctuary.

As they reach the sanctuary boundary, signs of human activity—rusted cans, plastic bags, and other detritus of past hunters and honey collectors—fade. Instead there are more signs of wildlife; the paw prints of wild dogs and leopards, a distant call from a barking deer, and bulldozed gaps in the forest where elephants have passed recently. The hunters are after smaller game; whatever bushmeat they can kill and carry out themselves and take back to their families. Anything that can't be eaten within a few days can be sold to the local market.

A few hundred meters ahead, a family of gibbons is feeding in the treetops—a mother and her six-month-old baby, and a father sitting nearby enjoying the morning sun. Since dawn the apes have been singing their haunting, whooping song in a duet that has lasted for the seven years they have been a tightly bonded pair. Their daily life is one of dawn choruses, swinging gracefully among the branches of the rain forest canopy fifty meters above the ground much of the day, finding ripe fruit to eat, and nurturing their infant.

Far below, on the floor of the forest, another inhabitant of the area is going about her morning. A forest tortoise, a brown boulder with armor-plated legs, is plodding across the leaf litter in search of rotting fruits fallen from the tree above. As the gibbons stuff figs into their mouths, they drop fruit that bounces near the tortoise; her breakfast. Her species is the largest in mainland Asia and can reach over two feet long and fifty kilograms, but she is only an adolescent of five kilos. She has lived in this patch of forest for eight years and with luck, she might live here fifty years more. Not far away are others of her species. Someday soon, males will find her and do battle among themselves over the right to court and mate her.

A blast shatters the calm morning, and as the male gibbon watches helplessly, his mate drops from her branch. She is lost from sight in the leafy canopy, only the cracking of branches tracing her plummet to the forest floor far below. A thousand birds and insects fall silent for a moment, then resume their day. The female hits the leaf-covered ground hard; she bounces once and flops over dead in a small pool of blood. A hunter walks slowly to her, kneels down and flips her body over to find the infant huddled to her chest, trembling but alive. He pulls the baby off its mother and holds it up to the sun, where it whimpers and struggles to free itself. The hunter slings the mother over his shoulder and stuffs the baby into a bag slung over his arm. He knows that city people will pay good money for a baby gibbon as a pet; it will entertain some family's children for a year or two. And the mother will go into the cook pot.

As he wipes the blood from his hands, his eyes catches something in the leaf litter. He leans over and picks up the tortoise. He has vivid memories of his father bringing large tortoises home to their village, where they would butcher them and eat the sweet meat. The empty shell would become a baby's bathtub or a serving bowl. In those days they were common; he would find them close to the village and play with them until they were killed and served for dinner. These days they're worth more sold to the local wild animal dealer. Wealthy foreigners, apparently with no concept of money or limits on throwing it away, will pay well for the creature in his hands. But he will keep this one for a family feast; or perhaps eat some and then sell the rest to one of his neighbors. He cradles the tortoise under his arm like a big loaf of bread and heads in the direction of his village. A gibbon and a tortoise for the pot, plus a baby gibbon for the animal dealers, make this a lucky day for him.

The Global Trade in Turtle Meat

The global trade in freshwater turtles is massive and dwarfs that of tortoises. Go to a food market in Beijing and stroll along the slatted wood walkways. There are fat catfish lying on mounds of ice on the left, piles of massive squid and clams on the right. There are rusted wire cages packed with wild-caught snakes and bins full of bullfrogs (American, farm-raised). Somewhere toward the front of the market are the turtles. There are hundreds of softshell turtles (mainly *Pelodiscus*), farm-raised in southern China or imported from farms in Thailand. And there are the ubiquitous red-eared sliders, America's pest turtle. Red-eared sliders are now a global species, like house sparrows or domestic pigeons. They have invaded every freshwater aquatic habitat. I've seen them paddling in swamps in Spain, rain-forest ponds in Thailand, and Buddhist temples in Myanmar. And in the market in Beijing, there are mountains of them, scrambling upon each other. They're farm-raised also, although a small percentage may be imported from the United States.

The global market in turtles increasingly comprises such ranched species. They're grown in huge outdoor pens in warm climates like Southeast Asia and fed commercial chows until they've grown large enough to be shipped to market. They must be tough, prolific, disease-resistant, fast-growing species. Tortoises don't fit this role well at all. They grow slowly, they don't lay many eggs per year, and many species cannot handle the crowded conditions of confinement. They are not good candidates for being raised as farm animals. So whereas many if not most of the turtles seen in food markets in China these days were farmed, any tortoise in any food market anywhere in the world is almost certain to have been taken from the wild. In most cases, these are illegal collections.

The turtle and tortoise extinction crisis in Asia therefore involves mostly turtles, not tortoises—entire nations have had their turtle fauna taken to supply the burgeoning consumer economy in China. It's been going on for thousands of years in all likelihood. Historical records show that turtle farming began in the nineteenth century in Japan and China, and such overcollection has probably changed the turtle and tortoise fauna in those regions since ancient times. But only in the late twentieth century did the human taste for turtles and tortoises become a mortal concern for them, and for us. In many of the countries involved, they're all on the menu: every species of turtle and tortoise is eaten, sold, and traded. China is not the only culprit. It's also about local consumption. Thais, Vietnamese, Cambodians, Laotians, and many other ethnic groups in the region love turtle and tortoise meat.

Chinese culture places a high value on turtles and tortoises. They are considered lucky, they are beloved as pets, and they are delicious eating. A recent survey in one area of rural China showed that 97 percent of people believe turtle meat is exceptionally healthy for the body. There are about thirty native turtle and tortoise species in China, and the Chinese have been eating them and grinding them up for traditional Chinese medicine for thousands of years. Ancient Chinese texts wax eloquent about recipes for roasting softshell turtles. As the Chinese diaspora has swollen, so has the pressure placed on fauna everywhere Chinese and other East Asian populations settle. Recently news reports bemoaned the loss of Tajikistan's native animal fauna, as imported Chinese workers created a market for turtles, snakes, and other game meat that led to rapid declines in native wildlife. A survey of turtles for sale in Chinese communities in California showed that the majority of nearly 3,000 turtles displayed in markets were almost certainly taken from the wild (because they were clearly too old to have been farm-raised). Most were the ubiquitous red-eared slider, but various other North American species were also for sale, all of them likely netted in a pond or lake in the United States. Even the highly endangered native western pond turtle (*Actinemys marmorata*) was being illegally offered. It had obviously been poached from the wild to sell in the market. A less direct impact of taking native turtles from the wild to sell in markets is that many well-meaning people buy turtles to release them in nearby ecosystems. These market turtles carry all manner of infectious diseases and stress-related pathogens picked up while piled in crates atop all the other turtles, and they then spread these pathogens into wild turtle populations.

And the Chinese trade is a two-way street. Millions of turtles are taken from North America to China each year; so many that despite all the turtle farms operating in China, it has become a net importer of freshwater turtles. From this we can assume that either domestic demand is increasing, domestic turtle resources are decreasing, or both. Although most are farm-reared red-eared sliders, some wild turtles from American ponds and streams are smuggled with them. Along with other wildlife smuggling, such as bear gallbladders, they make up part of a new and sinister wildlife trade that threatens to contribute to the problems many species already face in the United States. As China has emptied the turtle and tortoise fauna from the rest of Asia, they've turned to the North American continent. State legislatures have slowly begun to erect hurdles to collection in the form of catch limits. In Texas, banning commercial turtle harvesting altogether is under consideration. You can bet that, although this would be a great boon for turtles, it would also reveal a shocking level of poaching for the export market. Most of the land on which turtles live, however, is privately owned (ponds and streams on ranches and farms). Who can restrict ranchers and farmers from making a profit by selling turtles on their land to the international market? If managed properly, such practices might become sustainable. Currently, they aren't.

China has shown some determination to slow the flow of American turtles across their borders, and it has also banned the commercial export of all but the most abundant farmed species. Turtle farms have boomed, mass-producing American turtles for the Chinese food markets. But the sheer scale of population numbers in China means that the consumer population will outstrip the supply. What's more, turtle-meat connoisseurs prefer wild-caught meat to farm-raised, much as our western palates can easily tell a bite of wild Alaskan salmon from a fat and flaccid farm-raised one.

Eating Turtles in America

Americans certainly don't think of themselves as a turtle-eating people, and if the subject of a mass turtle and tortoise extinction due to the global food market came up in a discussion, few Americans would consider their culture to blame for it. But a century ago, America was the site of a turtle slaughter that rivaled that of the bison or passenger pigeon, driven by the American palate. The diamondback terrapin is little known today except to those who live along the Atlantic and Gulf coastlines between New England and Texas. It's a gorgeous, ornate big aquatic turtle, its head splashed with black spots on a field of pearly white, and its massive shining white jaws built for shattering crab and mollusk shells. On a recent spring morning I sat in a New Jersey salt marsh watching as the terrapins clambered out of a bay onto hillocks of grass, their clown-painted jaws and skin sparkling in the sun.

Such a scene is possible today, but the species appeared to be headed toward extinction less than a century ago.

Diamondbacks are notable as the only North American turtles that live in salty water—usually bays and estuaries where the salinity would kill other species. The species has evolved salt-removal glands that allows it to thrive in food-rich waters of the coastal shallows. It has successfully colonized habitats from Cape Cod to Texas, from Florida mangrove swamps to Chesapeake Bay.

The diamondback may be an evolutionary success story, but a century ago it was the focus of an industry. Wagonloads of turtles were sold on the streets of Manhattan, and as the taste for terrapin grew, so did the gourmet appeal of it, and the market value. Terrapin soup, as much a restaurant delicacy then as swordfish or sirloin is now, was sold in every fine restaurant. Wholesale prices of up to ten dollars per terrapin were not unheard of in the mid-1800s. Shipped by the barrel to Europe, terrapins began appearing on menus in Paris and London. They were diced into stews, boiled in soups, and roasted and eaten right out of the half-shell. In fact so many terrapins were transported around the coastlines of the United States that today the genetics of the species may be confused because terrapins from Florida occasionally ended up in New England and altered the natural gene pool.

Although some attempts were made to farm diamondback terrapins, nearly all on the market came from the wild. By the early 1900s, the plunder had peaked. The increasing rarity of the species and ensuing expense led to a major decline in the harvesting effort. There is still a market for terrapins today—thousands are sold in areas where Asian immigrant communities stoke the demand. And there is still a niche in some regions for the far more abundant snapping turtle as the prime ingredient in turtle soup.

The Insatiable Demand

So the insatiable demand for turtles and tortoises as food is not limited to nonwestern palates, or to the Far East. Because tortoises can't be farmed, any depredations on them for the Asian food market are unsustainable. The hundred thousand elongated tortoises reportedly taken from the wild for the food markets in a recent year in Asia cannot be replaced by captive farming. Turtles are netted, caught on fishing lines, and searched for in the muddy bottoms of swamps. Trained dogs are often used to locate forest species. A potentially more devastating part of the trade, but one not as visible, is the collection of millions of turtle eggs each year, which can quickly destroy whole populations even when the adults are not being caught in large numbers. In some cases nesting females themselves are the targets.

Some of the species most in local demand are the same species for which a lucrative pet smuggling industry exists. The little *Pyxis* spider tortoises are erased from many of their suitable habitats by collectors seeking them for the pot, not the pet store. Tiny though they are, they're the Cornish game hens of the region, on the menu for family feasts and sold on the black market not only as smuggled pets but as meals. Radiated tortoises are also victims of human hunger. Travelers on the dusty back roads of Madagascar have described local buses screeching to a halt so someone can leap off and pick up the beautiful creatures by the roadside. Despite their high value as illegally exported pets, such roadside collections are more often made for local consumption. Conservationists routinely report finding hunters' camps littered with the roasted remains of large numbers of radiated tortoises. They are taboo to eat for some local ethnic groups, but delicacies for others. They are on the menu of many eateries in southern Madagascar and are collected by the thousands for shipping to big city restaurants. When other foods are scarce, such as during drought years, the local endangered wildlife always suffers, and radiated tortoises are no exception. Even the local taboos normally in force can be lifted to allow locals to make use of a ready-in-the-shell food source. What can be done? In some regions, conservationists try to substitute easily grown, cheap foods for the rare wildlife, usually with limited success. Cultural traditions die hard. If you can't get people to stop eating tortoises, perhaps the trade can at least be regulated through better enforcement. In many of the same places where radiated and other tortoises are routinely eaten illegally, other animals that were once on the menu are not.

Beauty Contests

The wind just never stops blowing over Cap Sainte Marie, Madagascar. With a vast expanse of parched plains on the land side and nothing but ocean for the next two thousand miles on the ocean side, this is one of the most windswept spots on Earth. A broad white beach curves away below the point, atop which a carpet of white limestone gravel and rocks gives the place the look of a graveyard in a wind tunnel. It's a stunning, bleak place. And it is home to the world's highest density of the world's most beautiful tortoises. Here, where remoteness offers some protection from poaching, there are more than three thousand radiated tortoises per square kilometer. In the rainy season, it's hard to walk a hundred yards without seeing a few starburst shells ambling past. Radiated tortoises live in many other places in southern Madagascar, but nowhere else can a tourist visit with a certainty of seeing them, if not actually stumbling over them.

Its dramatic scenery and wildlife ought to bode well for Cap Sainte Marie and its inhabitants. Such a place, though currently difficult to visit and not developed for tourism, might twenty years from now be a hot-ticket destination. Tourism revenue could also be an incentive to declare more of the tortoises' habitat as nature reserve and to protect it via patrols and community outreach. Tourism in combination with development projects that raise living standards locally and reduce the pressure for taking radiated tortoises from the wild may help as well. But as we saw earlier, sustainable development is not necessarily sustainable. The market price of radiated tortoises is absurdly high. Some would argue that this is because international treaties forbid the buying and selling of these tortoises without permits that are very difficult to obtain. This makes them hard to acquire and so desired all the more. Poachers will sooner or later arrive in Cap Sainte Marie. When they do, they will, as they have in so many other places, rapidly leave a beautiful landscape devoid of anything but rocks and blue sky.

Madagascar is home to many species of lemurs, lower primates that are found only on Madagascar. Lemurs were once table fare for many Malagasy people, but no more. Although some poaching no doubt exists, for the most part the value of Madagascar wildlife to the nation's budding ecotourism industry has led to wider public awareness of the need to stop consuming the raw material that drives the tourism—beautiful endangered animals. Some small animals are so secretive and inconspicuous that tourists never see them, and so they cannot be used as flagships for that industry. With their beautiful starburst shells, radiated tortoises do not have that problem.

Such measures may keep the radiated tortoise from being eaten out of existence in the next twenty years. But rational, well-planned measures often go astray in places like Africa. The best plans work because of deep involvement and sense of ownership by local people, which involves changing the culture in a small way before the conservation picture can be improved. Such plans are easier to achieve when the lure of hard currency is the carrot. If Cap Sainte Marie and other tortoise strongholds were big-ticket ecotourist destinations, the government might see the full benefit of preventing poaching and protecting the land and its animals. Funds and infrastructure might then emerge. This is the pattern repeated throughout the developing world; if local people are given a reason to protect a resource, they will act rationally and do so, often passionately and always to the best of their abilities.

Stemming the Flow

One approach to preventing the tortoises and turtles of Asia from being eaten out of existence is preventing the vast illegal commerce in species bound for the food markets. For example, Indonesia is a major feeder country for turtles and tortoises bound for the stomachs of a billion Chinese. Bangladesh and Myanmar are the same. Some Asian countries don't want to have their entire native fauna shipped off to China. Others don't mind at all so long as government officials get their palms greased for enabling the trade. In Indonesia, for example, Sumatra has long served as a shipping point for turtles and tortoises being sent to China, Hong Kong, and Singapore. A study by Chris Shepherd of TRAFFIC (a nongovernmental organization that monitors the global trade in wildlife) found that more then twenty-five tons of tortoises and turtles were exported weekly from Sumatra. That's a hundred tons per month. And this figure includes only what is easily documented; the real number may be twice as high. Animals are trapped all over Indonesia and shipped to several distribution centers. The trappers use middlemen to get their wildlife to market. They're packed and warehoused and then sent by sea and air cargo to East Asian food markets. The meat of the Asian forest tortoise, *Manouria emys,* is particularly favored for its flavor. This large species is, along with the elongated tortoise, the primary tortoise victim in a trade dominated by vast quantities of freshwater turtles. A short distance away, another nation also has a booming trade in turtles, but with at least some regulations in place. In Malaysia, many of the same tortoises and turtles occur, and all are also eaten locally. But legislation exists that at least in theory protects chelonians, as a fisheries resource, from some of the uncontrolled mass collection they face elsewhere in Southeast Asia.

Given the staggering numbers, this trade won't last long. Unlike the collection of tortoises and turtles for the pet trade, which can devastate critically rare and sought-after species but on the whole is only moderate in its scope, collection for the food markets is wholesale and wanton. It's not a harvest; it's a rape, especially since the demand is booming at the same time the availability is plummeting. We're witnessing a carnage and mass extinction that will occur not over many decades, like the American bison or passenger pigeon, but over the next several years. Although certain species are favored by local people as the tastiest—softshell turtles—virtually every species has commercial value, and almost none can withstand wholesale human depredations for long. Some of the victims are fifty-pound river turtles; others are tiny forest tortoises. The food markets don't seem to know or care which species are critically endangered and which are not; all species turn up in rough proportion to their availability.

Farming to Extinction?

The holding ponds in Guangdong Province, China, stretch off in the morning mist for half a mile. What look like a million insects bobbing on the surface turn out to be turtles' heads, all turned to look at the intruders. Most of the turtles here are Chinese softshells *(Pelodiscus sinensis),* which are bred in the hundreds of millions by an estimated fifteen hundred turtle farms currently operating, legally or illegally, in China. But also present are thousands of turtles of other species, including some that are endangered in the wild (the Chinese three-striped box turtle, endangered and highly prized for its meat, exists in the hundreds of thousands on farms), and even including some that are native to North America and elsewhere. There's not much point in trying to get information from the owner of the farm about his operation. He's a businessman in a cutthroat billion-dollar industry that guards its secrets closely, and in a country where every venture happens on a vast scale, turtles have become very big business. Because of the level of secrecy, the scale of turtle farming in China has only recently come to light.

Chinese herpetologist Shi Haitao and his colleagues recently surveyed turtle farming practices here. Most of the farms are in the southern or central provinces, where warmer temperatures foster rapid growth to market size in these cold-blooded livestock. In addition, many farms are farther south in Taiwan, Thailand, Malaysia, and Indonesia, even though turtles raised there are nearly all shipped to China. Most species are being grown for the food market, with smaller numbers raised to be ground into traditional Chinese medicine or sold live into the pet market. Tiny Vietnamese leaf turtles are raised for the pet trade alongside abundant Reeves' turtles, right near American snapping turtles.

We don't have very precise numbers on the volume of turtles bred on Chinese turtle farms, but we can use figures from one of their American counterparts. Concordia Turtle Farm in Louisiana is the largest producer of turtle eggs and hatchlings in the United States. They sell fifty thousand young turtles of a variety of relatively common American species—sliders, cooters, painted and map turtles—into the American pet trade every year. These are first raised to the minimum four inches required by law to prevent the spread of turtle-borne salmonella virus (the thinking being that very small baby turtles are more likely to be groped, kissed, and even mouthed by children).

But that number is a minuscule fraction of the quantity of turtles produced. Concordia produces about 1.4 million turtle eggs per year, nearly all of these red-eared sliders. Of these, 85 percent hatch. The eggs are hand collected, with some thirty thousand per day dug up in the sand surrounding the many holding ponds. The million-plus hatchlings are headed for China, where they will be raised in the turtle farms. Others go to Korea to be used in Buddhist ceremonies that involve the release of turtles for luck. The hatchlings are sold for thirty-five cents. But these days, the vast network of Chinese farms are producing their own masses of hatchlings that cost only thirty cents apiece. Given the egg production at Concordia, it shouldn't surprise anyone to hear that Chinese farms hatch eggs from a variety of species in the hundreds of millions each year.

Turtle farming is good for turtles, right? You might assume it would take enormous pressure off rare species, allowing people to walk into a market to buy dinner or a pet instead of taking a wild animal. Conservationists are not so sure. First, the abundance of farmed turtles in China makes it difficult if not impossible to get an accurate idea of the real rarity of many species in the wild, because they are far more abundant on farms. It's also not clear what percentage of the endangered species being farmed are truly farmed, as opposed to being taken from the wild and then kept in holding pens until sold. Many "farms" use primarily wild animals as their breeding stock—these are known as "ranches" in the industry. They take wild animals, get them to reproduce in captivity, and sell the progeny to make their profit margin, without investing the years of care and expense to create self-sustaining captive populations.

Paradoxically, turtle farms are, according to some conservationists, a leading threat to the future of turtles and tortoises in China. Turtle farmers believe that each generation of captive-bred turtles are a bit less fertile than their parents were. So the farmers prefer wild-caught animals as their breeding stock. What is billed as farming is more often ranching. Wild turtles must be captured in enormous numbers to provide the breeding stock necessary to produce the best profit margin. While a few species, like the Chinese softshell, may be harvestable in sustainable numbers, most species are clearly not. Highly endangered species can only be profitably bred by removing many from the wild first, to build up a captive breeding stock worth investing time and money in. So the farms do far more harm than good for rare species. The main benefit of turtle farms is the money it puts in the pockets of the larger operators. Critically rare species are easily "laundered" through turtle farms, especially those species that are too difficult to breed in large scale in captivity. The existence of large farmed populations makes it easy for Chinese dealers to claim that wild-caught animals are bred in captivity. What's more, when individuals of rare species occasionally escape from farms and return to the wild, they may carry with them harmful pathogens picked up in the farms that could spread en masse through wild populations. This has certainly happened in other parts of the world, as we saw for desert tortoises in the American Southwest.

You could even make the argument that turtle farming is ecologically illogical. Although it's a protein-rich food for a developing nation, turtles are fed a high-protein diet to induce rapid growth to market size. In the final analysis, the protein that goes into the system may not be equal to the protein that comes out of it. So for the sake of slaking the Chinese hunger for one of their cultural dishes, the world stands to lose an entire order of vertebrates, and in a wasteful manner to boot.

In other words, although at first blush turtle farming

sounds like a great boon for relieving pressure on wild populations, it may actually greatly exacerbate it. To transform the industry into one that aids conservation efforts, the culture needs to change. Wild-caught turtles are still preferred for the dinner table over captive ones, and the trade in endangered turtles by farms needs to be regulated. It appears all but certain that someday soon turtles in China will exist only on farms and nowhere in the wild. There is a deep historical reverence for turtles in Chinese culture as symbols of luck and longevity. How tragically ironic that a civilization like China, with its overarching respect for its history and culture, would favor short-term profits over long-term management, and disdains the very concept of preserving biodiversity.

5

"SUCH HUGE DEFOURMED CREATURES"

Domino keeps coming at me. He noticed me sitting on the ground a few yards away, scratching one of his friends, and decided to get some face time. I have spent many years working with animals of all sizes and shapes, yet this is a unique experience. There is nothing like having a six-hundred-pound tortoise walk confidently into your lap. He moves with the steady determination of a rolling boulder, and as he reaches me he raises himself as high as he can on his four tree-trunk legs. He stretches his neck far and high, a chelonian Statue of Liberty. He's inviting me to pet him just as a cat arches her back and rubs against your leg. Only thing is, the cat's not going to crush your foot if you're not careful. I now have two tortoise heads, each the size of a small melon, suspended in the air in front of my face, both wanting some attention. All they want is to be scratched where it itches. I comply and stroke their necks, the backs of their legs, and that big deep triangle of leathery skin that forms a massive armpit. Domino looks blissfully at me as I rub him, his eyes huge, watery black discs.

I can't help but think of the story of the blind men and the elephant. Of course the dome of the shell would be a give-away—no other creature has anything so bizarre as that. But the blind man feeling the melon-sized head of this big male, and the one feeling his tail, and the one grasping his thick legs might not have any idea what sort of beast they were attempting to identify. He is of a scale we are not accustomed to wrapping our minds around.

Something else we have trouble wrapping our minds around is how much of the globe was populated with these behemoths so recently. We think of the giants of the Galápagos as mere curiosities, but are oblivious to their former glory. A few thousand years ago, they lived everywhere from Florida to Madagascar, from the Indian Ocean to the Caribbean. Giant tortoises were part of the landscape in many tropical ecosystems. They filled the role of grazing animals—deer, antelope, and allies—in tropical grasslands. Some weighed a half-ton. These reptilian rhinos lived until just the last few thousand years.

The global radiation of giants included the ancestors of our two remnant modern populations in the Pacific and Indian Oceans. We know very little about the direct ancestors of the Galápagos giants, except that they lived on the nearby mainland of South America. Genetic studies have recently shown that the closest living relative of the Galápagos giants is the nondescript little Chaco tortoise of Argentina and Paraguay. It's reasonable to assume, however, that the direct ancestors of those behemoths on the islands were large too. Some scientists believe that the Galápagos tortoises are actually dwarfed versions of their ancestors, since some fossils from the region indicate much larger tortoises in the recent past. If tortoises reached the islands the way all nonflying animals do— by floating across the sea on some flotsam, or in the case of tortoises, just bobbing in the waves until they hit a beach, then size matters. Large tortoises are far better able to survive many weeks at sea without drinking water and being battered by waves. Their lungs, located just below the dome of their carapace, provide some additional flotation assistance.

Halfway around the world, Aldabra tortoises are sometimes seen bobbing in the surf, having presumably gotten washed from the beach by an unexpected wave. Indeed, at least one record exists of a barnacle-encrusted giant tortoise plodding out of the surf onto a beach in East Africa. It likely had drifted from Aldabra, a tiny coral atoll in the Seychelles Islands. Depending on prevailing ocean currents, its journey may have lasted many weeks. The emaciated tortoise survived its long ocean voyage in a way that few other organisms possibly could.

Although this intrepid voyager probably started off in Aldabra, its ancestors made the journey in reverse. Giant tortoises living on the island of Madagascar off the eastern coast of Africa occasionally fell into rivers and were washed out to sea, hauling themselves ashore as far off in the Indian Ocean as the currents carried them. They landed in the Seychelles Islands, which include the tiny coral atoll of Aldabra. They also floated farther east and colonized the Mascarenes: Réunion, then larger Mauritius, and tiny Rodrigues to the east. Just as in the Galápagos, each island's tortoise evolved into distinct species (researchers disagree on whether each island has a separate species or should be considered subspecies of one umbrella species), each with a slightly different anatomy. On Rodrigues and Mauritius there were at least two species each, one larger and one much smaller.

Then the hand of man intervened. In 1690 François Leguat,

searching for a safe haven with his group of refugee Protestants cast out of papist France, headed for the Mascarene Islands aboard the sailing ship *La Hirondelle*. They aimed for Réunion, overshot it, and disembarked on Rodrigues. Leguat described a land carpeted in tortoises; the large species featured a saddle-shaped shell eerily reminiscent of those in the Galápagos. They were most abundant near the coast, living among the many palm trees and other locally endemic reptiles. The tortoises were so common the settlers had to wall off their gardens to prevent the crops from being gobbled.

They were even found on the many tiny islets dotting the lagoon just offshore. But once discovered by people, a small island is more a prison than a sanctuary. The vast herds became a walking steakhouse for the settlers, and once the word got out, a drive-through meat supply for any ship passing the island on a long ocean voyage. The same scene transpired on island after island: discover, marvel at the natural wonders, then turn them into takeaway dinners. But unlike the Galápagos, where a tightly clustered island archipelago sits six hundred miles from the coastline of South America, the Indian Ocean islands are scattered. To the north sit the Seychelles, the westernmost of which is the tiny islet of Aldabra. Hundreds of miles to the southeast are the Mascarenes. This distance made it harder to wipe out the fauna of every single island.

Some records of the plunder exist. Alexandre Pingré, the only trained naturalist to see the giant tortoises when their habitat still existed, was stranded on Rodrigues in 1761, at the tail end of the period of maximum exploitation. He estimated that at least 280,000 tortoises were taken from the island between 1735 and 1761. By the time Pingré arrived, the tortoises were too rare to be worth harvesting commercially, given the back-breaking work involved in carrying huge tortoises from remote parts of the island to the harbor. By the 1790s not only the tortoises but also the other endemic species of Rodrigues and neighboring Mauritius were dead forever. While the direct killing of tortoises was the main culprit, certainly the clearing of forest and grazing by livestock, slash-and-burn agriculture practiced by the locals, and the introduction of rats and cats that ate eggs and newly hatched tortoises played a role.

As sailing ships came and went, the crews collected vast amounts of tortoise meat. It was tasty, nutritious—the liver was especially savored—and highly transportable. It was canned food, alive. You could load a hundred tortoises in the hull of the ship, stack them on their backs like so many barrels of salted meat, and they would survive long enough that the crew could eat fresh meat for months. Soon only the most remote islands remained as natural tortoise refuges. Then even those were discovered. The whalers and early settlers hit the Galápagos with a vengeance, wiping out perhaps a half-million tortoises in eighty years. By the time Darwin landed there in 1835, the massacre was well under way, to be completed by the early twentieth century. What the whalers left, the earliest scientific expeditions to the islands took, bringing back the beasts only to watch them die an even slower death in a clammy London or New York menagerie.

Mauritius's native tortoises were among the casualties of the eighteenth century. But today Mauritius's offshore islands, as well as other islands in the Mascarene chain, have been repopulated with tortoises whose ancestors came from

Aldabra in the nearby Seychelles. During the nineteenth century even Aldabra, spared the slaughter for so long because it was far from the main shipping lanes, came under attack from the tortoise hunters. The original population size will never be known but was presumably in the hundreds of thousands. By the time it became clear that there would soon be no tortoises left at all in the whole of the Indian Ocean, the British scientific establishment made a plea to the British colonial government in the Seychelles. Take some tortoises from Aldabra, they argued, and relocate them to Mauritius for safekeeping, so that in the event of renewed slaughter or natural disaster there would be a second population. Charles Darwin himself, an elder statesman of British science at the time, cosigned the letter, and what became known as "Darwin's tortoises" arrived by boat in Mauritius, several hundred miles to the southeast.

There is a modern controversy raging about the tortoises of the Seychelles. For the century and a half since their scientific discovery, it was assumed that all the giant tortoises found in the Seychelles were imports brought from Aldabra by boat over the years. Then a Seychelles-born biologist, Justin Gerlach, took up a study of the giants. He noted that although nearly all the tortoises in the Seychelles were clearly Aldabran, there were a few here and there that had distinctly different shell shapes and other minor unique features. He studied every museum shell and skeleton he could find of some of the extinct Seychelles tortoises. He also played investigative journalist, searching backyards and private collections of tortoises in his home islands. And his persistence paid off. He turned up a number of tortoises that appeared to be not *Dipsochelys dussumieri*, the Aldabra giant tortoise, but *D. arnoldi* and *D. hololissa*, two other species believed long extinct.

Rediscovering any "extinct" species is thrilling. Usually the species is some small, secretive animal living in a patch of unexplored tropical forest. Rediscovering a gigantic tortoise that has been living in someone's backyard for the past century was, well, hard to believe. And after the initial excitement died down, a number of experts came forward to say that in fact they didn't believe it.

The two new species were discovered among tortoises that were already known to exist. They were incipient species—their identities reassigned, split from those of Aldabra. Redefining a species can be accomplished in two ways: anatomically, by identifying differences that make it clear the incipient species is really a distinct, new brand; and genetically, by seeing that the creature's DNA is clearly divergent from that of its former fellow. Gerlach had relied on his own anatomical evidence. He was able to gather up the remnant members of the alleged new species, do some captive breeding, and show that the newly hatched babies look somewhat different from those of Aldabran stock.

Gerlach's nonbelievers based their skepticism on both anatomical and genetic grounds. First, although both *D. hololissa* and *D. arnoldi* have distinctly different carapaces, tortoise shell growth is highly subject to the whims of nutrition, climate, and rearing conditions, not just genes. So had these individual tortoises that Gerlach considers to be new species simply been just Aldabran tortoises raised on a bad diet or cramped in tiny pens, they could well have ended up with deformed shells that resemble those of the extinct museum specimens.

Molecular genetic research has also cast doubt on the identity of the animals being attributed to the two extinct species. A research group headed by molecular biologist Jeremy Austin of the British Natural History Museum in London found little genetic difference among the putative new species. This may mean that they should be considered races of one wide-ranging species, just as most authorities consider the Galápagos tortoises to be one species. In all likelihood, the tortoises of the various islands of the Indian Ocean colonized them by bobbing across the waves from Madagascar, where we know a giant tortoise once lived. Alternatively, these so-called species may not be different in any meaningful way at all, and there is only one Indian Ocean tortoise today in place of the several well-differentiated forms of the past.

Rewilding in Mauritius

I am in the Mascarene Islands for my encounter with Domino, who may be one of the world's largest living tortoises. He lives at La Vanille, a sort of mini zoo and tortoise breeding facility set among sprawling sugarcane fields on the island of Mauritius, a thousand or so miles east of the eastern coast of Africa. Domino is believed to be in his nineties, although visitors to La Vanille seem to assume he is much older—the apocryphal centuries-old tortoise. An Australian biologist named Owen Griffiths has brought over a hundred of these giants to La Vanille, buying up tortoises from local pet owners, sugarcane plantations, and the like to create a huge breeding group.

Most are likely the grandchildren of those original Darwin's tortoises sent from Aldabra a hundred and thirty years ago. Their place in history doesn't matter to them as much as mealtime; at mid-morning they stampede like a herd of slow-motion buffalo to get to the mounds of veggies thrown to them.

And they do breed well. The saving trait of the Aldabra tortoise is that it is a hardy species—not a finicky eater or a sensitive patient like many of its tortoise kin—and also quite prolific. Griffiths is producing hundreds of baby Aldabra tortoises each year. Standing in the nursery, I'm surrounded by eight-ounce mini giants, and I try to imagine them when they've increased their bulk by some eight hundredfold. We have selectively bred dogs, cats, and birds for centuries. Recently, designer snakes in rainbow colors have become fashionable. But outsized reptilian pets are often a problem—unwanted pythons dumped into the Everglades are wreaking ecological havoc; brown snakes introduced to Guam have wiped out most of the native fauna. I can't get my mind around giant tortoises as backyard pets or as abandoned behemoths wandering in the local park.

But Griffiths has a different plan. He is attempting the rewilding of the Mascarene Islands. In the jargon of the field of restoration ecology, "rewilding" means attempting to recreate a semblance of a vanished ecosystem by re-establishing species in their former habitat. Ideally, rewilding includes not just a single species, but whatever species made the ecosystem run. When wolves were reintroduced to Yellowstone after a seventy-year absence, many other elements of the place began to work as Nature intended in the past. The wolves ate the elk, the elk stopped overeating the sapling trees, and co-

Older radiated tortoises are often suffused with a starburst of golden rays.

The Española Island race of the Galápagos tortoise has been brought back from the brink of extinction by captive breeding.

Lonesome George is the last survivor of the slaughter of Pinta Island tortoises. He has lived at the Darwin Research Station since the 1970s. Scientists are still hopeful of pairing him with a female of largely the same genetic ancestry.

These baby giants are Galápagos tortoises bred in captivity at the Darwin Research Station.

The Aldabra tortoise, the Indian Ocean analogue of the tortoises of the Galápagos, is being introduced to islands all over the Indian Ocean region where their extinct close kin used to thrive.

Aldabra tortoises are prolific when bred in captivity for reintroduction on other islands and for the pet trade.

In Kaeng Krachan National Park, Thailand, Pratyaporn Wanchai, a graduate student, studies the movements and behavior of the Asian forest tortoise.

This impressed tortoise has a transmitter attached. Radio telemetry allows scientists to track tortoise movements and study their diet and other behaviors. (Photo courtesy of Chey Koulang.)

Ecotourism may save some of the glamour tortoise species, such as this Galápagos tortoise on Santa Cruz Island.

Last of its kind: an adult male angonoka in the wild at Cape Sada, Madagascar. (Photo courtesy of Peter Paul van Dijk.)

A baby angonoka in the wild at Cape Sada, Madagascar. (Photo courtesy of Peter Paul van Dijk.)

Whether tortoises have a future on Earth will be determined
in the next decade.

than a rock, five hundred acres in size, sticking up out of the Indian Ocean. Access is completely restricted, and even if it weren't, its lack of proper boat-landing points maintains its isolation. To get onto the island you have to time the ocean's swell with the rocking of your craft and then make a nifty leap onto the rocks. Round Island is famed among biologists for its unique animals, all of them reptiles. Two lizards—Telfair's skink and Guenther's gecko—plus two unusual and primitive forms of boa, one of them possibly extinct, are found here and nowhere else.

Round Island also once had large tortoises, judging by old eyewitness accounts and by the remains of shells that have been found there. Today, Round Island is Mauritius in microcosm—its native plants and landscape thoroughly ravaged by the human introduction of rabbits and goats. Conservationists have spent the past two decades removing the pests, uprooting most of the nonnative plants, and replanting the place with native species. But if the rewilders are right, you need a big grazing animal to truly engineer the original landscape, by gobbling up the nonnative stuff and enabling native plants to grow. So a study will see if the introduction of proxy tortoises—Aldabras substituting for the larger species and radiateds for the smaller—would work. Christine Griffiths, a doctoral student at the University of Bristol (and no relation to Owen), is studying the feasibility of tortoise releases. She has established fenced-in plots on the island of varying sizes, with both tortoise species munching happily away all day on the grasses and other vegetation. If Griffiths is right, the tortoises will feed heavily on the introduced, undesired plants but mostly ignore the natives, allowing the natural ecosystem to slowly heal itself.

Why would the tortoises pass up the native plants? This would have to do with the plants themselves. Heterophylly is a well-known occurrence in plants in which the juvenile leaves have a completely different structure and appearance from the adult leaves. The local Mauritian ebony tree has large oval leaves, whereas its seedlings have long needle-shaped leaves in the tree's early years. The same tree may have adult-shaped leaves in the crown and juvenile-shaped leaves nearer ground level, giving it a grafted, hybrid look. These juvenile leaves are not only oddly shaped—they're also fairly indigestible to tortoises. If this pattern holds true for many of Round Island's plants, the tortoises should graze the nonnative species out of existence, allowing the native plants to retake the place.

But there's a third problem. There is precious little natural habitat left in the Mascarenes. They, like the Hawaiian Islands, were utterly transformed by the introduction of nonnative plants and animals that have driven the local varieties to widespread extinction. Mauritius, the largest island, is roughly the size of Maui in the Hawaiian chain but has a human population of 1.2 million (nearly ten times that of Maui) and is growing exponentially. Therefore, it may be most practical to rewild tortoises on offshore islands and the remaining patches of habitat on Rodrigues. Mauritius is ringed by a number of coral islets, some of which are government nature reserves. Some of these now function as natural laboratories, where animals that have no hope on the main island may reestablish themselves in splendid isolation.

This was originally the dream of the famed conservationist Carl Jones, a Welsh-born biologist who more than thirty years ago decided that Mauritian wildlife didn't all have to go the way of the dodo. The writer Douglas Adams wrote this about Jones's tenacious dedication to preserving the fauna: "his sheer perverse bloody-mindedness is the major thing that stands in the way of the almost total destruction of the ecology of Mauritius." He set about organizing the effort to save the last remnants of the original biodiversity of Mauritius, and became, along with the Mauritian Wildlife Foundation, the Saint Jude of the island's fauna. Today, the rewards of that labor include the pink pigeon, flapping around the nature reserve of Île aux Aigrettes (Egret Island) and the forests of Black River Gorge. They number about four hundred, or 390 more than there were in the 1970s. There is the Mauritius kestrel, a falcon often seen fluttering over the Mauritian landscape scanning for prey. (I saw my first one an hour after I arrived, not far from the airport.) In 1970, perhaps seven birds remained and they were nearly declared as good as gone. Today, there are perhaps a thousand roaming Mauritius. The list goes on: echo parakeets, the finchlike Mauritian fody, the warbler-resembling olive white-eye. All are breeding and rearing precious young at a sanctuary at Black River run by the Mauritian Wildlife Foundation. These are species for which conservationists had given up all hope. Today all are success stories, recovering in the wild.

Jones's approach is extremely concrete. His tactics for saving a critically endangered species might make many conservationists blanch. He believes in breeding animals in captivity as much as possible—restoring species by using these animals as the nucleus of a future population. He even advocates harvesting eggs from nests of rare species that will lay more eggs if they lose their first. The approach of some conservation organizations such as the Sierra Club is to protect the habitat;

they are loath to remove animals from the wild for any reason, even if reduced to just a remnant few. With the regulatory barriers and interagency infighting that mar conservation efforts in the United States and United Kingdom, half the species that Jones has saved in Mauritius might well have slid into extinction while awaiting bureaucratic hurdle-clearing in the west.

Thankfully, the Mauritian government allowed the conservation strategy to proceed. Île aux Aigrettes was one of the first sites chosen by Jones and Owen Griffiths for in-situ release and breeding of endangered Mauritian animals. It is a stone's throw from the mainland of Mauritius, the turquoise coral-dotted harbor at Mahebourg so near that at low tide you could just about walk to the place. I spent a day there visiting the biologists and their projects. Île aux Aigrettes is a coral dot in a large lagoon, nothing more, yet conservationists have morphed it into a nature reserve that hosts some of the rarest animals in the region, and that is also nearly free of invasive species. Eleven giant tortoises have been brought to the tiny islet and graze around the research station, with more planned in the future. Birds that could not survive a few hundred yards across the lagoon in the face of competition from more aggressive interlopers are doing just fine here on the island. Mauritian fodies and olive white-eyes flit through the trees. Pink pigeons are as easy to see on the island as domestic pigeons are on the streets of New York.

Most of all, it's the plant community that has changed dramatically on Île aux Aigrettes. After a massive effort to eradicate invasive alien plants—pulling out thousands of them by hand—the natives have retaken the island. The scrub that had covered the place for decades is being supplanted by a taller, more luxuriant growth, so much so that the observation platforms that used to look out across the island now have obstructed views. And many of the plants being restored are prime tortoise food. Vivid evidence that tortoises play a critical role in the local ecosystem is provided by large clumps of tortoise droppings that dot the place, each with a stand of ebony seedlings sprouting from them. It's possible that tortoises play an essential role as dispersers of seeds; they forage on leaves and fruit, and days later excrete seeds in their droppings far from the parent tree. The dung–as–seed nursery hypothesis is undergoing testing right now, as Christine Griffiths studies whether ebony and other tree seeds are more likely to sprout when sitting in tortoise dung or on bare ground.

By planting thousands of native trees and plants, Mauritian conservationists hope to restore tracts of forest that have not existed since the eighteenth century. The early results are promising. Île aux Aigrettes is barely recognizable to those who knew the place before it was restored. The canopy of the islet is growing higher and denser each year, and indications are that a larger tortoise population will clear out much undergrowth, recreating something closely resembling the original ecosystem. That, combined with the several nearly extinct animals that have been restored to the island, would make it a great and important, if somewhat unheralded, achievement in conservation.

A Safe Nursery in the Galápagos

The bus pulls off the paved road on Santa Cruz Island and we all hop out with our cameras and knapsacks, like ten-year-old kids on a class trip. It was a short ride from Puerta Ayora, the quaint harbor town that feels just barely south of the border to a visiting American, with its upscale gift shops, tee-shirt stands, and Mexican-American style cantinas. We drove up a winding road, passing from the prickly-pear dryness of the coast to a highland region where moss and moisture hang heavily on every tree. Our Ecuadorian guide leads us down a well-worn trail, through a picturesque ancient lava tube, and out the other end into a broad expanse of overgrown pastureland. There's a shaggy pony on the left, two skinny cows on the right, and nothing down the middle but Galápagos tortoises. *Chelonoidis nigra porteri,* precisely, the Santa Cruz Island race of the giant Galápagos tortoise. Unlike the bloated laggards I've seen in zoos, these tortoises are browsing actively across the wide pastures, looking like nothing more than a herd of cold-blooded cattle. At the first sight of the tourist mobs, they suck in their heads and their shells heave down mightily in place, smashing grass and flowers underneath.

Up close, a wild Galápagos tortoise looks quite different from its zoo cousin. First, these are all males, the females tending to remain lower down in the mountains to be nearer the beach for the nesting season. Males are enormously larger than females—these must be approaching five hundred pounds each. Their carapaces are blotchy gray and white—a moldy look that is a product of life in the highlands, where clouds and fog hang in the valleys and the humidity is nearly always high. And they have a look that says they've been working for their living instead of sitting around a zoo enclosure every day awaiting their tub of vegetables.

But I use the term "wild" advisedly. These giants are wild

in the sense that they are roaming free on Santa Cruz, unrestrained except by roads and other humanmade hazards and barriers. But they're in a cow pasture instead of anything closely resembling a truly natural habitat. This is private land, the landowner lucky enough to be ranching in a place where behemoth tortoises happen to live. In a rare case of ecotourism that works for both property holders and wildlife, the tortoises roam freely and the rancher charges admission for tourists to wander across his land looking at them.

It's not a perfect system. The island is rapidly being carved up into towns, villages, asphalt roads and the like. The tortoises just want to keep doing what they've been doing for millennia, but they often can't. The Santa Cruz Island subspecies of Galápagos tortoises is in better shape than many of the others—three of the fourteen known varieties are extinct, with a couple more teetering on the brink. There are some three thousand left. But relative to the species as a whole, the Santa Cruz tortoises may survive just because they are valued by local landowners.

Trying to escape the tourists, I wander off toward a boulder and sit under a distant shade tree. Of course the boulder turns out to be a tortoise, waiting out the morning human intrusions in the shade. His head sunken deep within the recess of his shell, he looks as happy to see me as his ancestors were to greet the whaling ships. My only affront will be to sit a few feet in front of his face snapping photos, and he dutifully cooperates. The whalers were not so benign. The eastern Pacific Ocean held great whale populations in the eighteenth century. The age of sail took awhile to reach the region, but once it did, the ships came and took whales until there were too few to bother. Herman Melville was a visitor on a whaling ship in 1837. He thought the "Enchanted Isles" as bleak as nearly everyone else of his day. But Melville's ship didn't come to study the tortoises. They came to collect them, in the same way that whalers and travelers in the far-off Indian Ocean had already collected nearly every giant tortoise from the Mascarenes and Aldabra. A few decades later, only the furnacelike, rugged conditions of some of the islands preserved some subspecies.

The Galápagos Islands and its giant tortoises have been written about so many times that superlatives have become clichés. "Living fossil," "majestic giants," and "the land that time forgot" are just a few of the eponyms given the place and its inhabitants. In fact, none are true. The islands are young and dynamic. Older islands at the eastern end of the chain have already slipped beneath the sea as new ones emerge far to the west. As they inch eastward on their tectonic plates, the islands pass over a hotspot in Earth's crust, giving the younger islands their signature volcanic essence. The western islands of the chain are so young that some of their animal inhabitants—including the tortoises—are older than their island homes. They've descended from lineages that existed on former islands when their current abodes were still beneath the waves. The tortoises themselves are hardly living fossils, despite their appearance. They are descended from tortoises of the mainland of South America and have been on their own evolutionary paths for as little as 6 million years. Although all their living kin are far smaller than they are, it seems likely that their immediate ancestors were large enough to have bobbed across the waves for weeks without dying. Although they may reach extreme ages well over a century in captivity, the rigors of life in the wild limit them to perhaps half that or a little more, despite their resemblance to Methuselah.

For me, getting to know the Galápagos and its tortoises has brought a number of surprises. Not all the Galápagos Islands have tortoise populations—some were apparently never colonized, and a couple of others have lost their tortoises to human depredations—but the tortoises on those that do vary tremendously in shape and size. There are two general shape categories, domed and saddle-backed, with gradations in between. The domed morph is the classic image of a tortoise; a carapace that looks like a well-risen loaf of bread. Of the eleven surviving varieties of Galápagos tortoises, six are dome-shelled. They tend to be very large—like those on Santa Cruz. They also tend to be found on islands that are lusher, greener, and rainier. In some cases, they are found up in the humid zones of the highlands that are lush above a certain elevation but remain dry along the coast.

The saddle-backed tortoise populations, on the other hand, tend to occur on islands or in habitats that are more arid and brushy. In such places, the leaves and fruits for which the tortoises spend their slow lives foraging are not abundant at ground level because of the furnacelike aridity. The prevailing notion is that on islands lacking lush ground plants, the tortoises' ancestors needed to reach up to browse from low bushes and tree limbs. Natural selection favored those with shells that upturned in front, leading over time to populations with a saddle-like shape to the shell. At the same point, natural selection appears to have produced prickly pear cacti that grow as trees, with trunks that protect the plant from the ravages of tortoises. One of the old Spanish names for tortoises, *galápagos* (which has replaced the original Islas Encantadas as the sobriquet for the island group), means a kind of saddle. The saddle-shelled varieties have a bit more room in the front of the shell to stretch their long necks upward. Although Darwin is often credited with observing the different morphs of tortoises and realizing the significance, it was actually the British vice-governor of the islands during the time of Darwin's visit, Nicholas Lawson, who passed on this observation to him. In fact, even the shells of the many tortoises brought on board the *Beagle* to be eaten on the homeward voyage were simply dumped into the sea rather than kept as valuable museum specimens. The three little tortoises that Darwin brought back to London alive were too young to show the typical adult shell shape.

Another distinctive feature of the Galápagos tortoises that stands out when seeing them in their natural environment in such quantity is how utterly different males and females are. There is no mistaking sex differences among adult Galápagos tortoises. The male is endowed with a long, fat tail, all the better to extend under the female's carapace to reach her cloaca during sex. In a large male giant tortoise, the tail and the impressively large pink organ stored inside its base are the length of my forearm and much thicker around. In addition, males have a plastron undershell that is so deeply concave that peeking under the shell when the animal is raised up high on his legs gives me a view of a vast inverted bowl. You don't keep your head under such a massive animal for long to get a good look, and flipping him over is not an option either. The plastron of a male is well dished out to allow him to mount a female and get his tail close enough to reach that desired cloacal opening.

The most salient male-female difference in these giants,

however, is sheer size. It's not easy to estimate the weight of a large tortoise, because having a rigid outer frame conceals what is inside. An emaciated, sick tortoise may show other signs of illness, but its shell looks just like that of a same-sized animal that weighs hundreds of pounds more. A healthy, fully adult male Galápagos tortoise will weigh, depending on the island variety, anywhere from 350 to 500 pounds. Females of the same race are little more than half that weight. It is painful to watch a male giant walk up behind a female, clamber onto her carapace, and attempt to mate while grunting, mouth agape—she is almost invisible under him. With his massive weight on her body during his prolonged sessions, she must be uncomfortable or worse.

We believe the extreme size difference between male and female Galápagos tortoises is the result of the Darwinian principle of sexual selection. When males compete for females, size matters in determining which male wins. Males battle over females, their elephantine necks stretched to the maximum and their mouths wide open in one another's faces. In many smaller tortoise species, males try to flip over one another, and a rival left helpless in the tropical sun won't live long. There's not so much physical combat in the Galápagos (and very little at all in their Aldabran counterparts), but females don't randomly mate with every male that comes along. Male size and strength are two key determinants of genetic quality and therefore make a good basis for selection when females are mating.

Among the domed varieties of Galápagos tortoises, males are enormous and females are merely huge. But the saddle-backed races, perhaps due to their food-poor diet in arid regions, tend to be quite small compared to the domed races. So small that if you saw a female of one of the saddle-backed morphs, you would not consider her a giant at all. She is almost the size of a backyard tortoise.

Leaving the Santa Cruz giants grazing in their pasture, we stroll back down the hills to Puerta Ayora. Huge, brightly painted sculptures of iguanas and gulls decorate the town center. This is a place that knows how to market itself. There are more gringos than Ecuadorians on the main street that follows the shore of Academy Bay. We stroll a mile from town, past art galleries (mostly reptile-themed), bars full of vacationing Americans, and a precious few fishing boats and remnants of local culture. At the end of the road is the Charles Darwin Research Station, home to the main breeding center for local animals, including Galápagos tortoises. There is another, smaller facility on San Cristóbal Island. The center has the look and feel of a zoo combined with a museum, which I suppose it is. Behind chain-link security fences, enclosures house newly hatched tortoises from several of the islands, including this one. I'm surprised that I can distinguish the populations even as hatchlings—some with flattened carapaces, other domed like a Volkswagen Beetle®. These pint-sized versions of their behemoth parents are already sparring and clambering about on lava rocks in their enclosures, which are intended to give them the feel for the islands that will someday be their homes.

The principle behind captive breeding is simple. Although the Ecuadorian government has captured, shot, and trapped every nonnative pest they could over the past couple of decades, many islands still have too many rats, and some still have goats. Goats nibble away every living thing, but the rats are mainly deadly to baby tortoises—a quarter-pound, soft-shelled hatchling is no match for a rat's long incisors and eat-anything mentality. So eggs are hatched at the research station, and the babies are given a head start here for several years. Once they are judged large enough to be impervious to rats—at around four to five years old and eight or ten pounds—they are taken in batches to their island of ancestral origin and released. Their survival odds once back in the wild are high so long as they get a safe start in the secure confines of captivity.

The pens are separated according to subspecies of baby tortoise, and each tortoise is painted with a color-coded number to indicate year of hatching and island identity. This program has had great success with at least one population. The tortoises of Española, formerly known as Hood Island, were essentially gone from the wild in the 1970s, reduced to two males and eleven females. Marksmen brought in to eliminate the problem goats did their job, but the tortoise population showed no signs of recovery. Brought to the Darwin station, they were later joined by a third male that had been a long-time resident of the San Diego Zoo. They began to produce the first baby Española tortoises the world had seen in a long time. Since then, more than a thousand Española tortoises have been hatched at the Darwin station, and many have been returned to Española.

Española is flattish, rock-strewn, cactus-covered, and generally a barren, hostile place. In the center of the islet rise low hills that provide a measure of cool humidity from trapped moisture. It looks for all the world like some desolate stretch of the coast of Baja California, south of the border from southern California. The tour boat pushes onto the white sand beach of a brilliant blue lagoon. We step across the line of sea lions lying on the sand like sentries and walk the craggy path around a section of the little island. Off in the distance are the hills where tortoises once again roam. This is the recipe for success in the Galápagos; research combined with ecotourism, scientific know-how abetted by a supportive local government.

Down at the Darwin Research Station I'm standing on an elevated boardwalk looking into a pen at the rarest animal on Earth. He's a species of one, staring back at me, his neck outstretched on the ground, his ink-dark eyes gazing up. Jorge Solitario—Lonesome George—is the last known member of the Pinta Island subspecies of Galápagos tortoise. He has become an icon of extinction, like the dodo of Mauritius or Ishi, the last Yaki Indian. Is he forlorn, aware of his lonely fate? In fact, he's probably just sleepy in the midday heat.

In 1971, seventy years after the last tortoise on Pinta was thought to have paid the butcher's bill to the whaling ships, biologists conducting research on Pinta saw Lonesome George ambling across the dry landscape. Not realizing the significance of this sighting, a year passed before another biologist learned of it and recognized its importance. Multiple expeditions were mounted on Pinta to search for George. A large tortoise on a small island might not seem a difficult target, but only through much luck was George found. He was transported to the Darwin station by boat, and has remained there ever since.

In tortoise years, Lonesome George is still a prime-age breeding male, but his time is running rapidly out. Despite more than one further search of Pinta, no more living tortoises have been found, and it's reasonable to assume that there are none left there. For years he shared his pen with two females from the slopes of Wolf Volcano in the northernmost part of Isabela Island. But George showed no interest in having sex with them. Even the tried and true manual ejaculation technique that tortoise biologists often use (palpate the area about the cloaca, usually producing an erection and ejaculation in no time) didn't work.

But even though there doesn't seem to be another living tortoise on Pinta, George's story has recently grown more intriguing, and perhaps less poignantly solitary. As the gull flies, Wolf Volcano is just about the nearest point of land to Pinta Island, so it makes sense that, lacking any true kin, conservationists paired him up with females from that part of Isabela Island. But George's kinship situation is not nearly so straightforward. Although the most recent expedition to Pinta didn't find any living tortoises, they did find fifteen skeletons of long-dead tortoises, all of them male. Yale University geneticist Adalgisa Caccone and her colleagues were able to extract DNA from these skeletons and others that were already in museum collections. As a result, the researchers now had a referent library of sorts, telling them what the genetic makeup of a Pinta tortoise should look like. They intended to use the data to identify any long-lost Pinta tortoises in zoos around the world.

Then something unexpected and exciting happened. Caccone's research team found a partial DNA match for George among tortoises living on Wolf Volcano on Isabela Island. Wolf Volcano has two separate populations of *Chelonoidis becki,* the local subspecies of tortoise. This population, known as the Puerto Bravo tortoises, is unusual among Galápagos tortoises in that instead of being either domed or saddlebacked, they exhibit a spectrum of shell shapes, with much individual variation. This was a tip-off; the diversity of shell shapes reflects underlying genetic diversity and a likely complicated history for the Puerto Bravo tortoises. DNA testing revealed that eight of the twenty-seven Puerto Bravo tortoises sampled were closely related to tortoises from Española, all the way across the Galápagos archipelago from Isabela. Furthermore, one of the twenty-seven appeared to be a hybrid between a Wolf Volcano female and an Española-esque Pinta male.

Having already discovered one hybrid that included Pinta Island genes in the first small batch of twenty-seven tortoises tested, the odds seem good that once the entire population of several thousand Wolf Volcano tortoises has been tested, more Pinta hybrids are going to be found. At the least, it would allow breeding projects that would, after decades of careful selective breeding, restore the Pinta tortoises. The question that remains is how Wolf Volcano ended up with a tortoise whose father hailed from Pinta, almost fifty miles away. The islands are connected by a prevailing ocean current that runs southwest from Pinta toward Isabela. The possibility of a few Pinta tortoises wading into the surf and being carried across the waves to Wolf Volcano is not so far-fetched. Another scenario for Pinta tortoises reaching Isabela is that ships may have thrown tortoises overboard, or stopped for provisions on Isabela and off-loaded Pinta tortoises, some of which wandered away. Further progress in genetic studies like Caccone's will answer questions about Galápagos tortoise ancestry and also create some new questions. How many other islands are harboring fugitive tortoise genes? How did they get there, and what will conservationists do if they determine that a substantial portion of any island's gene pool hails from other islands?

The ultimate goal of the Darwin Research Station is to restore the Pinta tortoises to Pinta. If that involves Lonesome
George, so much the better. After being paired with new part-
ners, George has fathered eggs, all of them so far infertile. But
whether the future father of Pinta Island tortoises is George
or a relative on Wolf Volcano, it is beginning to seem a real
possibility that the island will be repopulated. And what's
more, genetic studies of the Galápagos tortoises may reveal
hidden hybrids on other islands in the chain. For example,
Caccone's team has discovered genetic evidence that the
Santa Cruz Island tortoises should be considered three distinct subspecies. One of these new, yet unnamed lineages exists in one tiny population of one hundred tortoises. The extinct tortoises on Floreana, so distinct that they may be a
separate species from the rest, actually have close genetic
matches still living on Isabela Island. In other words, they
may not be extinct after all, exactly. One potential obstacle is
the recent finding by some of the same Yale University researchers who discovered the hidden variation among the
tortoises that what we today generally consider subspecies
perhaps should be considered full and distinct species. This
finding, if widely accepted, would argue against mixing genetic lineages from different populations. Observing hidden
variation at the genetic level informs our historical under-
standing of the tortoises' history, but it also raises questions
about conservationists' ability to accommodate these new
forms in their future plans.

Lands of the Giants

The island giants of Aldabra and Galápagos have parallel histories: evolution from a mainland ancestor and adaptation as

a dominant island herbivore, reaching enormous population sizes in small areas. Then discovery by humans was followed by wholesale slaughter, with a last few preserved, and finally intense conservation efforts have restored minimum numbers. But these two tortoise populations have quite different present situations, and possibly very different futures. Aldabra atoll is completely isolated today; access to its beaches and lagoons is limited to a few scientific trips each year. Barring a natural disaster such as global warming could bring, with Aldabra rising only a few feet above sea level, that population of nearly 100,000 tortoises is safe. Moreover, Aldabrans have been introduced widely in the Indian Ocean, with many more introductions planned. It seems that a few decades from now, the Aldabra tortoise will exist in the hundreds of thousands again, not only on Aldabra atoll but scattered across the Indian Ocean and, if some conservationists have their way, farther away too.

When we try to reconstruct ecosystems and restore endangered species to them, we make a choice about the time period to which we are trying to restore the ecosystem. The California condor made a last stand in southern California before the last birds were captured, then captive bred, their progeny now being released. But two hundred years ago the condors had roamed over a far wider area in California, and a thousand years ago—just a blip in evolutionary time—they had occurred all across the southern United States, as far east as Florida. So should we ultimately seek to restore them to their entire ancient range, or just to that tiny piece of the range that remains in human memory? A small but vocal school of thought in conservation would apply Owen Griffiths' vision of Aldabra tortoises as proxy species more widely. Thus, since some of the Caribbean Islands were home to giant tortoises several thousand years ago, one could make a case for ecosystem health there being dependent on a large grazer like a giant tortoise. Aldabras or Galaps would fit the bill quite well. Other regions of the world where people drove giant tortoises into extinction, like Madagascar and parts of mainland Africa, might benefit from their reintroduction as well.

If Aldabra tortoises could serve as ecosystem engineers on islands that are near their evolutionary epicenter, why could they not prosper in more far-flung places as well? Wouldn't ecotourists in the Caribbean love to encounter behemoth tortoises while vacationing? How about stumbling across a Galápagos tortoise while hiking on one of the Hawaiian volcanoes, or watching them wade through waterholes in East Africa? The idea is heretical to traditional conservationists, but so would have been campaigning for California condors to soar above the Grand Canyon, where they have flown for years now. Introducing endangered species into available ecological niches in climatically favorable places where their close kin once lived may be the face of twenty-first-century megafauna conservation. Giant tortoises may be merely leading the pack.

The reason this idea has taken hold among biologists studying Aldabra tortoises but not yet for Galápagos tortoises is that the Aldabran species is far more abundant. If and when Galápagos populations reach the thousands for each subspecies, introductions elsewhere will likely become a reality for them too. And you may then stumble on a huge tortoise almost anywhere in the tropical parts of the globe.

If that is to happen, the Galápagos ecosystem will have to be preserved. Few tourists cruising the islands in their luxury tour boats realize that as recently as 2000, local fishermen rioted, angry that they were not allowed to harvest lobster past their government-mandated quotas. Because of the mushrooming level of immigration by Ecuadorians to the islands, more fishermen means the overall fishing limits are reached sooner each year. The Darwin Research Station was attacked and ransacked, staff were evacuated under threat of violence, and at one point the rioters issued an ultimatum: they would behead Lonesome George if the quota was not lifted. Ultimately, the government struck a devil's bargain, raising the annual quota of lobster to a level that most ecologists believe is unsustainable in the long term, and which will only lead to depleted lobster populations, ever more fishermen, and more violence.

These concerns are far from the minds of most Galápagos ecotourists. The most visible problem for tourists is the crush of tour boats trying to land at the same wildlife-viewing spots all at the same time. Like so many other places on Earth, what brings tourist dollars to the Galápagos economy also threatens to destroy the place. Meanwhile, the tortoises keep lumbering about their ancient lands, unconcernedly eating, sleeping, and mating.

6

BELOVED CAPTIVES

On a steaming hot day in Chatuchak Market, Bangkok, Thailand, the crowds arrive early, and by nine in the morning it's hard to walk without stepping on feet and bumping into people a foot shorter than I am. Chatuchak is one of the world's biggest flea markets, covering an area the size of several city blocks. Everything you can imagine and thousands of items you can't are for sale. The smell of Thai stir-fry from the ubiquitous vendors permeates everything. Tourists and locals mingle. The tourists search hundreds of aisles for wood carvings, silk, and other exotica for their living rooms back home. The locals come to shop for household wares, knock-off designer jeans, and good, cheap foods.

In addition to the acres of stalls and shops selling their wares, there are rows and rows of live animal dealers. Many of these sell tropical fish. Hundred-dollar koi loll in massive plastic tanks. The eponymous Siamese fighting fish circle in their tiny bowls. It looks like a huge pet store without a roof. But there are other animals too. Interspersed with the stalls selling puppies and kittens of every breed—or near miss—and color are wild animals for sale. These days they are mostly pedestrian creatures like squirrels. One shop has a bin of baby flying squirrels with a hastily crafted cardboard sign warning against the use of video cameras. The owner knows it's not legal to sell wildlife in Thailand. Some years ago undercover video showed all manner of endangered animals being sold, from parrots to gibbons. If it wasn't openly on display, you just had to ask and someone would lead you furtively to a back room where a deal could be brokered for a leopard cub. Although leopards may be a rarity in Chatuchak these days, otters, lorises, monkeys, and a host of exotic birds are for sale if you ask the right vendor.

Down the aisle from the squirrel vendor are several tiny pet shops. They would not look out of place in any western country; tidy and smelling pleasantly of animal food and waste. In the storefront windows there are tortoises, but not the red-footed and spurred and Russian tortoises found in every western pet shop. There are radiated tortoises, star tortoises, pancake tortoises. These are tortoises that would not be found in any western shop because of the price they would command—thousands of dollars for the radiated—and because under normal circumstances they cannot be sold in the United States.

In March 2008 police raided Chatuchak and found twenty radiated tortoises for sale. The radiated tortoise is a CITES Appendix I species, meaning that the Convention on the International Trade in Endangered Species, the international body created to prioritize species for protection based on their status, has placed it on its list of the most endangered, Appendix I. This listing carries concomitant restrictions on commerce and transport. Nearly all other tortoises and turtles are Appendix II, meaning they are threatened but not on the brink of extinction, and can be bought or sold across international boundaries with proper import and export permits. Besides protecting flora and fauna from being removed from their natural habitat, CITES listings are also intended to raise awareness of the endangered status of animal species to the government of the countries in which they live, and hopefully to push that government to support action to save them.

Right now, however, I'm in an open-air Thai market where the rules don't apply. The authorities do occasionally inspect these shops, but enforcement is difficult. I ask the owner of one shop, a portly, well-spoken young guy, about his radiated tortoises. They are, he assures me as he lifts one gently up to display it, captive bred in Thailand, meaning they were not taken from the wild. But of course this is a lie. They were likely smuggled out of their native Madagascar. From there they would have gone to Singapore—a black-market hub for the international wildlife smuggling trade—and on to Bangkok and other Asian cities. Or perhaps they were captive bred in one of those places, from parents that were smuggled out. I point to the Indian star tortoises in the next cage. Unlike the radiateds, stars are not yet on the CITES Appendix I, although they are rapidly disappearing from their native Indian subcontinent. These, he says again, were captive bred in Thailand. Unlike the radiateds, the stars are mostly adults, making it extremely likely that they were smuggled from India, where a huge industry thrives on poaching star tortoises. The same raid that confiscated hundreds of radiated tortoises also found three ploughshare tortoises, or angonokas, for sale. There are perhaps six or eight hundred angonokas left on Earth. So a measurable fraction of the entire species were available as a smuggled commodity in Chatuchak on the day of that raid.

How the animals are smuggled is up to the ingenuity of the smuggler and the ineptness of the customs inspection process. Recently a government raid on a home in a cozy suburb of Los Angeles led to the arrest of a Hong Kong national who had been smuggling Indian star tortoises into the country from Singapore. His accomplice in Singapore presumably had the wherewithal to get the tortoises out of India without being caught. The tortoises were then packed in boxes labeled "action figures" and shipped to Los Angeles International Airport. The top layer of the shipping crates had toys. Underneath, packed like eggs in cartons, were many dozens of tortoises, each worth several hundred dollars on the open market in the United States. Western Europe, North America, and Japan have long been the destination for smuggled exotic animals. Recently the Chinese, who consider turtles to bring good luck, have joined the ranks of consumers willing to spend vast sums and break the law to obtain their prize. With a burgeoning market and ineffective enforcement of trade laws, smuggled species can't withstand the onslaught much longer.

It might sound like hyperbole to suggest that the global trade in tortoises and turtles rivals the international narcotics trade. But evidence suggests that along certain routes, it's true. It's lucrative, easy, and lacks the taint of deep immorality that drug trafficking carries with it. If you can smuggle drugs across borders, you can smuggle all but the largest tortoises. They tolerate awful packing conditions for long periods, they don't need food or water during the smuggling operation, and they are easily laundered through other countries or states.

The Desire to Acquire

The fate of endangered species would be vastly different if humans did not have such a fetish for feeling connected to animals. Some do it by hunting and killing animals—which

seems a bizarre sort of connection to me. Others do it by pho-
tographing animals, or by hiking or driving to see wild animals. Unfortunately for many species, others connect with
animals by owning them. Pet ownership can provide compan-
ionship and satisfy a sense of wonder about the natural world.
Certainly some animals—dogs and cats—are loved because of
the unconditional affection they give their owners. In western culture they have become largely surrogate children. But
other pets are craved without any real (though perhaps some
falsely perceived) return emotional attachment. Sometimes
they are loved as mere ornaments of great beauty—they are
breathing baubles.

Why are people attracted to tortoises? They certainly em-
body many attractive qualities. They are ancient, dinosaurian.
They have the wizened face of an old philosopher. They are
so utterly harmless and so benevolent in demeanor that they
evoke a desire to protect and nurture. Some species do relish
human contact—although most do not—and will come when
they see you, seeking food or a good scratching of the neck.
They are vegans. They don't make a mess, they don't smell,
and they can go many days without eating or just feed happily
on your lawn grass. And they are often stunningly beautiful,
huge biological gemstones that can walk across your backyard. They are, in a word, cool.

Having spent a week sitting among Aldabra giants, I can at-
test to the connection. It's hard not to feel something for an
animal that is so large and antediluvian and has the sweetly
grizzled face of that benevolent movie alien ET. Their desire
to get into your lap to be scratched endlessly is one of the most
endearing traits I can imagine in a reptile. To be looked in the
eye by such a creature is to look into an old soul.

Some cultures have other reasons for keeping pet tortoises

and turtles. In China, when they are not being eaten, they're kept as pets for their alleged good luck and the prospect of longevity. Buddhist temples across Asia keep turtles and tortoises in their ponds, usually with good intentions but to the detriment of the animals. I once saw an elongated tortoise—a forest species—bobbing around in a stagnant green pond at a temple in Yangon, Myanmar, that was as alien to its natural environment as an astronaut on the moon.

For these reasons and probably many more, people have kept tortoises as pets for centuries. Because tortoises are tough animals, they often survive abysmally inhumane conditions for years, succumbing only when the cumulative effects of malnutrition, poor sanitation, and temperatures far too cold or too hot for them take their eventual toll. And then the grieving pet owner may just go out and buy another tortoise.

The most pernicious aspect of the live tortoise trade is that the universal law of collection obsession applies. As with rare stamps, rare coins, and rare everything else, tortoise species are desired in direct proportion to their rarity. A beautifully patterned tortoise that is abundant in Nature will be largely ignored by collectors until it becomes rare, at which point it becomes much more desirable. Ironically, once a tortoise is placed on Appendix I of the CITES treaty, making it harder to obtain or transport, it takes on a legendary quality and fuels a desire to own one despite the price tag that goes with it. Radiated tortoises are beautiful, but certainly no more so than a number of other turtles and tortoises—eastern box turtles rival them in sheer color intensity and personality. But box turtles are still common in many wooded areas in the United States, whereas radiated tortoises, found only in a small area of Madagascar, are being poached out of existence by local people for both food and the pet trade.

So a box turtle commands a few dollars on the open market, and a serious collector would turn up his nose at devoting time to breed them. But a radiated tortoise can command a price of thousands of dollars by virtue of being difficult to obtain. This perverse equation of scarcity, beauty, and dollar value is hell on tortoises. It's nearly impossible to make accurate estimates of the scale of the black market, but certainly there are many critically endangered species sitting in private collections in the United States, Europe, and the Far East. In most cases these animals were smuggled out of the wild. The collectors, however strongly they profess their love of the animal and desire to protect it, support and perpetuate the illegal trade as surely as an alcoholic supports the liquor business that he may publicly rail against.

As a creature becomes rarer and extinction looms, regulatory protection is enacted and the black-market value skyrockets, along with the demand by collectors with deep pockets. When a species is moved from CITES II (can be sold across U.S. state borders and internationally) to CITES I (cannot be sold interstate or internationally without federal government permits requiring a lengthy application process), the monetary value of the species soars.

Educated, otherwise enlightened people who support wildlife conservation become predators on the last remaining wild tortoises in the name of acquiring captives. These folks usually justify their avarice by arguing that they want to breed the species in captivity "before it goes extinct in the wild." The problem is that for the vast majority of species, they are better off right where they are than in someone's backyard.

For a few critically endangered species whose habitat is nearly gone—think angonokas or Galápagos or Burmese star tortoises—captive breeding by experts in a dedicated research facility can make a difference by creating new captive populations that can eventually be returned to the wild if and when effective habitat protection is enacted. But for every dedicated private collector and breeder who may contribute something to the captive population of a tortoise, there are ten who will not. For those tortoises, captivity is genetic death, the end of their lineage. The aim of CITES restrictions is both to hinder commerce in these species and also keep them out of the hands of all but zoo professionals.

Yet the federal government's well-intentioned bureaucratic hurdles to possessing or selling such rarities is so full of loopholes and unintended results that it sometimes seems the laws are designed to drive the species into extinction. There are critically endangered species that have no captive-bred population precisely because bringing a group together to establish a colony for long-term preservation is nigh impossible due to regulatory barriers. And the CITES designation of Appendix I, while in theory based on the latest census data from the wild, is often subjective and irrational. Radiated tortoises are believed to number in the millions in certain protected habitats in Madagascar. They are listed in Appendix I because of the rapid rate at which local people are collecting them, both for food and for sale to the black-market pet trade. But scarce they are not. In fact they are almost the box turtles of Madagascar, seen in markets and as backyard pets, and they have made their way as pets to the islands of the Indian Ocean. Meanwhile other tortoise species still await listing as Appendix I despite being in terrible trouble in the wild; the delay is often due to politics or the competing interests of a variety of species all worthy of listing.

Another quality that can doom a tortoise is beauty. Nothing turns a collector's head like beauty, and coupled with rarity, it's a toxic combination. This is the dilemma of all the starred tortoises, from the Indian star tortoises to the South African geometric tortoise to Madagascar's radiated tortoise. Their shells make them irresistible. Size matters too; the giant species are always in demand, and even the most closely protected populations are smuggled. Thieves have raided breeding facilities of the giants and made off with hatchlings, each worth thousands of dollars on the black market.

So a lethal system has evolved in which the tortoises with the most precarious population levels and highest sensitivity to poaching are also the ones most likely to be highly desired and therefore smuggled. As rare tortoises have increased in market value, the risk of being caught has been outweighed by the profits to be made. Tortoises are easy to steal: they tend to be kept outdoors and if you know where they live, they are easily carried off. They can be kept in appalling conditions for weeks without any external sign of poor health. So they are hidden in crates of toys, buried in suitcases, and carried in pockets, like so many diamonds. Stealing a few radiated tortoises is like committing grand auto theft in the potential payoff, if not the penalty if caught. Owners and zoos often resort to microchip identification implants, although that won't help unless the tortoise is recovered.

But the pet trade in many species does not need to depend on backyard thieves. Enough of Madagascar's tiny spider tortoises of the genus *Pyxis* can be found among the leaf litter on the forest floor in the right habitat of southwestern Madagascar to earn a month's wages in a morning. Although spider tortoises are also collected for food, they are pretty enough to command high prices in the global pet market. Protected areas and nature reserves are not safe havens here. Madagascar continued to allow exports into the pet trade for years after the two spider tortoise species were supposed to be fully protected. As tortoises with limited reproductive potential—*Pyxis* lay an egg or two at a time and a modest total for their yearly breeding efforts—they cannot possibly withstand intense collecting. They have rapidly disappeared from areas where just a few years earlier they had been common as rocks.

Since the 1980s, a new and particularly sinister aspect of the desire to acquire has threatened tortoises. Since the late 1980s, thirteen new species of Chinese turtles have been described, leading many of the world's foremost turtle and tortoise authorities to declare that this extraordinary biodiversity must be protected. However, most of these new species have never been seen in the wild, and repeated attempts to trace the specimens to any natural point of origin have failed. All the while, the black market demand for the new species skyrocketed. Any newly named species becomes the latest fashion to be acquired at all costs.

This led James Parham and Jonathan Fong, doctoral students at the University of California, Berkeley, along with renowned Chinese turtle biologist Shi Haitao, to undertake a genetic study of the newly described species. They found strong evidence that the thirteen "new species" are in fact artificially produced hybrids of two existing species. In some cases, they are hybrids of distantly related species not even belonging to the same genus. But if the resulting progeny are distinctive looking and attractive, they are a highly market-able commodity. Apparently, depending on whose version of the facts you accept, even some high-profile western turtle and tortoise scientists and collectors were duped into supporting and unwittingly subsidizing the export from China of these false species. The massive turtle breeding farms of southern China are the likely culprits—entrepreneurial spirit combined with a healthy profit margin on the hybrids led to their production. Meanwhile, precious conservation energy and money was spent searching for and trying to breed these sham species. Since so few specimens of some of the newly proclaimed species exist, it is impossible to say whether the hybrids have resulted from a few lucky crossings or a systematic attempt to dupe the buying public into believing in the existence of new, undescribed species. The issue is not fully resolved—there are experts on the taxa in question who challenge the hoax proponents. Some of those experts have a vested stake in the faux species, however, having been led down a garden path in naming them and declaring them valid. But it is an extraordinary example of the kinds of conservation problems the unceasing pressure of the pet trade can bring to bear on turtles and tortoises.

Bizarre Is Best

Some species are highly sought after because of their oddity or uniqueness rather than their beauty. Pancake tortoises are not especially attractive and are rarely eaten by local Africans, but they make fine pets, small enough to be kept in apartments. Moreover, they have that wonderfully eccentric flattened version of a normal tortoise shell. They are highly adapted to life in a crack; the crevices in rocky outcroppings on the East African savannah are their habitat. Outside the crevice, their pliable shell and diminutive size make them vulnerable to a host of predators. But once they scamper back into the rock, they wedge themselves into a tiny space where they are safe and sound. Except from poachers. Pancake tortoises are pried out of their homes with crowbars so often that some rock outcroppings no longer have crevices for future generations of tortoises. Since rock damage is permanent, whole outcroppings are depleted not only of this but of all future generations of the little tortoises. Their reproductive output is low: the female's flattened body can only hold one or two eggs at a time. They cannot withstand the current onslaught on their numbers in the wild. For years they were imported in moderate numbers; the importing country simply required a certificate showing that they were legally exported. Such certificates were easily obtained or forged. The government of Tanzania—home of most of the world's remaining pancake tortoises—wisely imposed an export ban in 1992. (Kenya had imposed its own export ban in 1981.) The price of the tortoises skyrocketed, and captive-bred babies became highly sought after, bringing more than five hundred dollars for a tortoise the size of a quarter. Meanwhile, although the European Union had also banned the import of pancakes, many European countries allowed importers to bring them in.

Then, early in the new millennium, the Tanzanian government briefly changed course on the zero-export issue, and suddenly pancakes flowed into the global pet market again. They were advertised for sale as "captive-hatched" or "legal imports," but typically they were neither. They had been jackhammered from their homes and smuggled out of Africa, with some exporters and local government officials in cahoots, no doubt lining their pockets on the backs of thousands of the little tortoises. Many of the imports were listed on import paperwork as coming from countries in Africa where pancake tortoises do not even live. Because the pet market was suddenly flooded with imported wild pancakes, the market price dropped dramatically, and casual pet owners instead of serious collectors and breeders could now afford them. Their newfound popularity led to even more demand for imports. Meanwhile, those private breeders who had worked for years to create a captive-bred pancake population to relieve pressure on wild ones no longer had a ready market for their baby tortoises, since they had been undercut by all the imports.

Turtle biologist Michael Klemens did some research into the international market for pancake tortoises in the 1990s before the export trade was initially shut down. Local Africans—the folks who break the rock ledges apart to collect them—earned as little as a nickel per tortoise. A local broker arranged the sale of quantities of tortoises to dealers for about forty cents per animal. By the time the exporter—the key player in that he has to obtain the permits to ship them out of Africa—got his hands on them, the tortoises were worth about fourteen dollars each. By the time they reached American pet stores, their sale value had increased to 150 dollars, a 3,000 percent mark-up.

Today pancake tortoises are available on the internet with a few mouse clicks and a credit card. Many of these are captive-bred babies of the imports, but in recent years there has been a spate of animals advertised for sale, nearly all of which have illegally entered the United States. Yes, the importers may have seen legal paperwork, but it's widely known that the species is not supposed to be reaching Europe and the United States at all, just as Mediterranean tortoises aren't supposed to be arriving in western Europe. Yet even as I write this, there are advertisements on the internet for "farm-raised South African pancake tortoises." Such offers may be a thinly veiled effort to legalize illegally smuggled tortoises. They've traveled from Kenya or Tanzania in someone's car trunk or suitcase into South Africa, whereupon they are slapped with the "farm-raised" label—exceedingly difficult to disprove—and then exported with legal paperwork from South Africa to your living room.

The First Pet Tortoise

Keeping tortoises as pets is not a recent fad. In the beginning, there was Timothy. Timothy was a small tortoise, the first documented case of the international trade in tortoises and the first pet tortoise to have its life history recorded in detail until its death. Beginning in at least the 1600s it became fashionable in Britain to keep a tortoise in the yard. Despite the long winters and often miserably cold summers, tortoises from the Mediterranean lived for decades under such conditions. They were often believed to have outrageously exaggerated ages, probably because they often outlived their original owners and so their actual age was forgotten.

Timothy (who we now know was female) was a specimen of *Testudo graeca*, brought from the Mediterranean to England in 1740 by a sailor and purchased by a Mrs. Rebecca Snooke, who cared for the tortoise until her death in 1780. Her nephew, a quiet clergyman named Gilbert White, dug the tortoise out of hibernation and took it back to his vicarage, where it lived until the year after White's death in 1793. White was a serious amateur naturalist in the village of Selborne. Timothy was given the run of the garden and lived there under White's watchful eye. White kept a journal of his observations of the natural world around him, which was eventually published as *The Natural History of Selborne* and ranks today as one of the most detailed and insightful early modern accounts of animal life. White measured Timothy (six pounds and fully grown, perhaps even elderly, when she came to Selborne), tested her hearing (by yelling at her), her swimming ability (by dunking her), her hibernation patterns, and her diet. Among the many other aspects of the natural world in the village of Selborne, Timothy was the only individual creature to receive the long-term focus of the naturalist's pen.

Timothy's fate was far better than that of the estimated 10 million other Mediterranean tortoises that have been taken from the wild and brought to Europe over the last century. She survived. Most, piled like rocks in bins for weeks while awaiting shipment following their collection, then packed like oranges in crates for shipping, do not. Between 1969 and 1972 alone, nearly a half-million "Greek" tortoises (actually hailing mainly from Tunisia) were imported to the United Kingdom. Of the tortoises that arrive in Europe healthy enough to be sold in a pet shop, 80 percent die within their first year in the hands of well-intentioned but incapable owners. By the third year out of the wild, an estimated 92 percent of all imported turtles and tortoises are dead. And this does not include the millions that were turned into curios, ashtrays, and other objets d'art for sale to tourists in the bazaars of northern Africa. Once tortoises along the North African coast of the Mediterranean began to become thin on the ground, the trade turned its attention to Eastern Europe, where marginated tortoises *(Testudo marginata)* and Herrmann's tortoises *(T. herrmanni)* could still be easily gotten.

Finally, in the 1980s, over two hundred years after Timothy, the United Kingdom banned the import of most Mediterranean species. This both ended one prime source for the international legal and illegal trade and also encouraged more serious attempts at captive breeding as a way to relieve pressure on wild populations. But the story doesn't end there. Following the break-up of the Soviet Union, a massive trade began in the little tortoises native to the southern and western nations of the former union. These "Russian" tortoises *(Testudo* or *Agrionemys horsfieldii)* are as hardy as any tortoise in captivity. It's a good thing they are: it has been estimated that hundreds of thousands entered North America and Europe over the past decade. The availability is high and there are few regulatory barriers to cleaning them out of their native range for sale overseas. The animals are cheap to buy in pet shops and hardy enough to withstand the shipping process. In short, there is thus far little incentive for anyone to breed the species in captivity on a large enough scale to replace wild-caught stocks.

More sinister is the recent opening of a gargantuan loophole in the import laws that CITES was created to close. Although wild-caught tortoises cannot be legally imported or sold, tortoises born or bred in captivity can. Great Britain banned the import of Russian tortoises in 1999, after which importers turned their attention to potential loopholes in the laws restricting commerce in wild-caught tortoises. All a dishonest animal dealer has to do is lie convincingly about the origin of his animals, and he may receive an import license. Animals that are sick, parasite-covered, and emaciated sometimes receive designations as captive bred when they arrive at Heathrow Airport. The best tip-off for many potential buyers is price; a captive-bred animal's price reflects the months or years of time invested in maintaining the parent stock and the time invested in breeding. The price of imported tortoises is always very cheap compared with those bred in captivity. Virtually every adult tortoise that you've ever seen languishing in a pet store was taken from the wild, regardless of what the shop owner assures you. So the trade continues.

It doesn't even matter to the pet industry whether the species in question can survive in captivity. The hingeback tortoises of Africa are still frequently, and legally, imported to the United States. These are tortoises with exacting climatic requirements—they need high humidity, warmth, and low light levels to mimic their rain-forest floor habitats. They are most often seen sitting in the harsh light of a flood lamp in an empty, dry cage in a pet shop. Their life expectancy in such conditions is a few months; being tortoises they die as slowly as they live. What's more, imported animals arrive with their digestive systems packed with parasites, having lived cheek to jowl with hundreds of other tortoises under highly stressed conditions in storage bins and shipping crates for weeks. The naïve buyer who takes one from a pet store has acquired a new pet that is already quite sick.

Timothy was, in light of all this, extraordinarily lucky. But like nearly all the other imported tortoises, Timothy didn't reproduce in captivity, and so from a genetic standpoint, she was dead the moment someone picked her up in Tunisia or Algeria and shipped her to a foreign shore. The vast pet trade wouldn't be so sinister if most of the animals taken from the wild were maintained in good health for many years and also bred, reducing the demand for wild-caught animals. But with the exception of a few species and a tiny fraction of tortoise collectors, this is not the way the pet trade works. Reptilian and amphibian pets are, for the most part, disposable commodities, bought in bulk by retail stores and sold as though they were inanimate objects. You can walk into the nearest pet store, whether a locally owned shop that has been on your corner for forty years or a huge chain store with too much inventory and too few knowledgeable staff, and see the same scene—animals kept in conditions that would kill a warm-blooded creature in no time. But these are reptiles; they're tough and durable. Live slow, die slow. The vast majority of pet tortoises die within months after their owners bring them home, and the owners attribute the death to neglect on their part rather than to the often parasite-loaded condition of the animal at the time of purchase.

In recent years an increasing portion of sales of tortoises and turtles has shifted to massive pet expositions, usually held at a county fairground or a convention center. Here vendors rent tables to display and sell their animals. The vendors are not all the same. Breeders have invested years of hard work to ensure they are producing healthy animals. Dealers, who often pay rent for their retail outlets back home, must pay closer attention to the bottom line. But dealers are bigger and sell more animals than breeders. Although signs at such expos advertise "captive-bred only," that is quite often not the case. Reptile pet expos are good venues for connecting buyers with sellers and provide a marketplace where comparison shopping is possible. But they are the puppy mills of the tor-

toise world nonetheless, churning out animals to deal to cus-
tomers who may be utterly clueless about how to take care of
their impulse purchase. And although the animals shown for
sale are usually (but not always) bred in captivity, the breeding stock that produced them was most often taken from the
wild. Ask any vendor at a pet expo about the impact he is making on wild populations, and he will proudly (and a bit defen-
sively) tell you that captive breeding is the solution to the
extinction crisis, and he is just doing his part. He'll say that
his goal is to reduce pressure on wild populations through
captive breeding. But he needs wild-caught tortoises as badly
as anyone does; without them, he would have to wait for
years for his captive-bred tortoise to mature (at a glacial pace)
enough to breed and produce the retail stock he's selling.

Some tortoises are easy to take care of and some are posi-
tively not. No one has figured out yet how to reliably keep and
breed the diminutive padloper tortoises of southern Africa in
captivity. They ought to be easy keepers, like the small Mediterranean tortoises, but they're not. They have a strong tendency to starve themselves when taken from the wild, and
even when they do eat readily, they're prone to dying suddenly anyway. The beautiful impressed tortoise (*Manouria
impressa*) from the hill forests of southern Asia is even less
of a survivor. The percentage of impressed tortoises that are
alive one year after importation is close to zero. Apparently
cursed with an exceptional vulnerability to internal parasites,
it succumbs when packed in bins and crates of other tortoises,
and by the time it ends up in an importer's hands, it's way too
late without very savvy medical intervention. Other species
are hardy enough to withstand much abuse and still arrive in
the destination country relatively unscathed. Unfortunately,

these tend to be the same species that are priced low enough to be affordable to anyone with a cardboard box and a head of lettuce, so their survival rate is still appallingly low.

Captive-bred tortoises are not the answer to the extinction crisis in most cases, in part because we know so little about where captive tortoises actually originated. Tortoise laundering is widespread, even the norm, in the wildlife import industry. Animals that are exported out of countries in which they do not even naturally occur make a mockery of the regulatory treaties. When pressed, animal dealers often cannot or will not tell you exactly where their imported goods came from, and their replies are often expedient or outright lies. That tortoise that was legally exported from Malaysia had been smuggled across the border from Thailand, where it had been secretly collected and furtively transported by truck, car, or ox-cart to some portion of the porous international border. Such transportation of tortoises makes it extremely difficult to accurately re-establish wild populations, except for those species whose distributions are so limited that there can be no mistaking their natural range. Putting a tortoise that hailed from a mountainous area of Cambodia into a low-lying part of Malaysia often means death for that animal. Even if it survives, it won't have the right environmental cues to reproduce.

And then there is the internet. There are Web sites devoted to the marketplace of reptiles, both captive bred and wild caught. Buying a tortoise on the internet is about as time consuming and requires the same degree of expertise about proper care as downloading the latest popular music. A glance at one of the most widely used sites for reptile sales shows that about a quarter to a third of the tortoise advertisements feature animals that have been taken from the wild, plus others that are coy about it but probably sell imported animals too. In all likelihood most of the remaining reptiles for sale, though advertised as bred in captivity, were produced from imported, wild parents. So there is good reason to be skeptical of the claim that tortoise collectors are reducing the demand for wild specimens. And anyone who uses the internet to buy or sell knows that claims can be especially arbitrary when business is not conducted face to face—all the more so when import paperwork is required. There are import and export quotas for most species, and many breeders are ethical when it comes to dealing in legally acquired animals. But exceptions abound.

The international trade in exotic animals is not conducted solely between the third world and the first world. Box turtles from the United States were still shipped by the tens of thousands each year to European countries, where they are considered among the most beautiful of all pet turtles, as recently as the mid-1990s. Field studies have shown that box turtles do not locate mates once their population density is reduced below a moderate level, meaning that removal of even just a large fraction of the turtles in a given area spells doom for the whole population. Many U.S. states have now banned the sale or export of their native turtles, but an international trade still exists in those that have not.

Ultimate Fates

So tortoises in the twenty-first century find themselves caught up in a consumer-driven market that knows few boundaries. Where is it leading? In a best-case scenario, the captive breeding of tortoises by private individuals would lead to a reduction in the pressure to import them and would relieve one gigantic pressure on those wild populations. Perhaps someday captive-bred tortoises could be reintroduced wholesale back to their species' original habitat. Species that are relatively hardy and prolific would become commonplace as pets, and then no one would care about buying imports. Imported tortoises are cheaper, but they are also riddled with parasites, acquired both in Nature and in transit when they are packed together with many other tortoises. Stress levels lower their immune systems' natural tolerance for parasite load. Imported tortoises also tend to be shy, having lived a life happily free of humans, whereas a tortoise hatched in captivity knows no other life and tends to be more phlegmatic, more personable, and an all-around better pet.

The ultimate pet turtle is the red-eared slider, ubiquitous not only to pet shops but to ponds, lakes, and streams the world over, and native to the southern midwestern United States. I've seen them swimming in lakes on the coast of Spain and jungle ponds in a Thai rain forest, and stacked like pancakes in a market in Beijing. If the red-eared slider is the ultimate pet turtle, the ultimate pet tortoise is the African spurred tortoise, which has replaced the time-honored Mediterranean tortoises because of the former's equal hardiness and affability, and far greater size. Often known by its zoological name, *sulcata,* the spurred tortoise is ridiculously tough and easily fed—it's a glutton—tolerates a wide temperature range, has an outgoing personality, and is easily and prolifically bred. Native to arid, hot lands along the southern edge of the Sahara Desert in western and central Africa, the *sulcata* is as abundant in American backyards as it is uncommon in its native range.

Why backyards? Spurred tortoises grow up quickly. The adorable little guy in the aquarium tank on your dresser quickly becomes a foot-long powerful grazer who bangs against any barrier placed in his path until it gives way. A few years after that he's a sixty-pound brute who can knock down picket fences, bully every other animal in the yard, and eat you out of house and home. Unwanted pets have existed since the beginning of domestication, but unwanted or abandoned giant tortoises are a new phenomenon. There are now rescue and adoption programs for these tortoises and many others. As if the spurred tortoise weren't too big for a suburban backyard, a growing number of giant Aldabra and Galápagos tortoises are being bred and kept as pets. For the price of a pedigreed dog, you can have a tortoise that, while it won't ever crawl onto your bed to sleep with you, will come for food, beg to be scratched on the chin, and live about ten times longer than the dog. But what happens when the adorable little football-sized Aldabra becomes a three-hundred-pound behemoth that can't even be given away?

There are animal rescue organizations these days for rottweilers, Maine coon cats, cockatoos, and all manner of high-priced animals. Their abandonment or maltreatment speaks volumes about the extent to which North Americans see pets as disposable. And there is also tortoise rescue. Tortoise rescue and adoption has gone mainstream in the past few years, especially in the southwestern United States. The desert tortoise, as its lands have been eaten up by housing developments, shopping malls, and off-road vehicular onslaught, has become a waif in its own ancestral home. A few thousand lucky ones each year, among the many thousands plowed under or crushed by cars, find themselves in adoption centers run by local volunteers. Some time after that they end up as backyard tortoises, where they may live happily for decades.

If the owner has a male-female pair, tortoises may breed, and many suburbanites in southern California have multiple generations of home-grown desert tortoises. But they cannot be returned to the wild easily, as I discuss in Chapter 3, and there are local regulations about keeping desert tortoises. In theory one must obtain permission to adopt a desert tortoise, have one's yard certified as tortoise-safe, and agree never to breed the animals. But there are far more unregulated adoptees in most communities than there are official adoptions. Fortunately, there are official adoption networks for potential tortoise owners that provide information on care. Such networks also erect some useful barriers for those who would casually pick up a tortoise in the desert and deposit it in their backyard.

7

ARE THERE SOLUTIONS?

The only way to get to Cape Sada is by passing boat or hired plane. It's a patch of scrubby deciduous forest on a small peninsula on the northwestern coast of Madagascar. The Baly Bay region is the last refuge of Madagascar's largest living tortoise, which also happens to be one of the rarest tortoises on Earth. The angonoka is not a living jewel like a radiated or star tortoise. It is mostly mud brown, with subtly contrasting squares of yellow within each shell plate. With its carapace as domed as a globe and a protruding bowsprit under the male's chin—hence the English name "ploughshare"—the angonoka is as distinctive looking as a mud-brown tortoise can be.

The angonoka exists in a wild population of perhaps two hundred animals. Once upon a time they occurred all along the coast. For centuries they were carried away by the thousands to nearby islands, where their meat was considered a delicacy. Later, their habitat was whittled away by fires that were set to promote the growth of pastureland for cattle. When the cattle had overgrazed the land and moved on, the habitat that recovered was too sparse to provide adequate shade and shelter in such a blazing hot place. These last pathetic remnants of angonoka habitat are now under intense threat from every possible source. Introduced wild pigs root up their nests and eat eggs and hatchlings. Local people have long valued angonokas more as pets than as wild neighbors.

But in the twenty-first century, these threats have taken a back seat to poaching for the foreign pet trade. As their rarity has increased, the value of the angonoka on the black market has skyrocketed. A local poacher will gladly risk six months in prison in exchange for one angonoka sale. He won't get the tens of thousands of dollars the middleman may ask for, but a hundred dollars to a village poacher is dinner for months for his kids. Go on the internet and search for "angonoka" and you will see videos of smuggled baby angonokas sitting in plastic take-out containers in kitchens in Tokyo. These were likely progeny of the wave of thefts that occurred at breeding facilities in the 1990s.

In 1986 Project Angonoka was founded with the goal of both protecting the tiny remaining wild habitat and captive breeding the tortoises. Those bred in captivity would be sent to Baly Bay to beef up the wild population. A captive breeding center was established at Ampijoroa in Ankarafantsika National Park, a few hundred kilometers from the home of the last wild population. By the mid-1990s the project had established a small breeding group of formerly wild (confiscated pet) angonokas, refined some techniques for getting them to reproduce, and reaped a bounty of some 140 baby angonokas. Then disaster struck. One night poachers cut through a wire mesh fence and made off with two adult females and seventy-four captive-bred babies. Before long the smuggled animals were reported to be in the Netherlands, from where they were presumably smuggled to buyers in Europe, North America, and Japan.

Although a few of the animals were later recovered, most were lost from the breeding project for good. But the work continued, and today hundreds more angonokas have been produced, with some released back at Cape Sada to boost the wild stock. It is a precarious success—new generations of angonoka offspring are protected by thick wire mesh and guard dogs—but the reward is a potential future for the species. The worldwide captive population of angonokas is minuscule compared to many other critically rare species, some of which are more abundant in private collections than they are in the wild. Only a handful—not counting the uncountable black-market imports—have found their way into western zoos and private collections. Habitat, of course, will never return once lost to development, so the angonoka will never regain more than a sliver of its former realm. But the captive breeding success achieved so far may well allow the largest, rarest tortoise in that part of the world to see the twenty-second century and beyond.

Throughout the first six chapters, I've painted a very bleak picture of the future of one of our planet's major animal groups—one of imminent extinction. But there are a few rays of hope. As always, both extinction and preservation from extinction are human issues, not animal issues. Without our impact on the Earth there would be no twenty-first century mass extinction looming. And now that it is upon us, we are the only salvation for those same victims.

The keys to angonoka survival in Madagascar have been twofold. First, foreign conservationists had to take an active interest and infuse the project with cash and leadership. This happened in the 1980s, when the Durrell Wildlife Conservation Trust, the brainchild of famed naturalist and author Gerald Durrell and the Jersey Zoo in the United Kingdom, set up the captive breeding center at Ampijoroa and at their home facility on the Isle of Jersey. They began to breed the animals in captivity and in the process trained local people who, instead of becoming tortoise poachers as they might have been in generations past, became impassioned angonoka advocates.

The rescue of the angonoka may be a budding success story, but can its lessons be applied elsewhere? There are two fundamentally different aspects to any onsite conservation effort, which must be somehow linked to attain any success. The biological side is obvious. We have to know how many tortoises we are trying to save, understand their habitat requirements, and respond to the threats to those habitat needs. Biologists spend years on shoestring budgets trying to fill in this missing information. Sometimes the plan calls for passive conservation; that is, learn what the animals need and then work to prevent threats to it. Sometimes, however, we engage in active conservation. Perhaps the animals have been reduced to small, fragmented populations so scattered that there is no hope of recovery unless they are brought together to form a new and viable unit of the species. Statistical models that predict the viability of populations can provide some guidance in establishing how big a population is needed, or how much continuous, nondegraded habitat is required, but none of that can be known until basic aspects of the species biology are understood. Some tortoise species lay one or two eggs a year while others lay fifty. Some take only a few years to reach sexual maturity while others take decades. Our database of long-term detailed information is extremely poor for the vast majority of tortoises and turtles, so even the most basic of field studies is vital for many species.

The second key aspect to any field conservation effort must deal with the cultural and political factors. Do local people respect the tortoise? Do they eat it? Do they simply plow its habitat under in the interest of making farms and villages? Or do they actively attempt to rid themselves of the animal, through collecting for the commercial pet or food trade? This is where biologists often drop the conservation ball, and where grassroots work from anthropologists or just dedicated volunteers can make an enormous difference in how the local culture views tortoises. Having a doctoral degree is not necessary here, and sometimes gets in the way. What is badly needed is feet on the ground, connecting with local people and, even better, training local people ("capacity building," in conservation jargon) to help themselves by reinstilling the virtues of wise, low-impact resource management.

At the extreme other end of Madagascar from the home of the angonoka, radiated tortoises are benefiting from local involvement in their plight. Most of the ancestral lands of the radiated tortoise also happen to be the ancestral lands of the Mahafaly and Antandroy ethnic groups of southwestern Madagascar. Of course the tortoise's ancestors lived there a few million years before people did, but the people always win in these sorts of land conflicts. These ethnic groups do respect the tortoise highly, however. The concept of *fady*, a taboo against harming an animal, is widespread on the island, and radiated tortoises are *fady* to the Mahafaly and the Antandroy. For generations the tortoises and many other local animals were protected by the *fady*, even revered. But times have changed, and along with shrinking land in an unforgiving landscape of poor soils, erosion, and severe deforestation, the *fady* has faded too. Younger people don't respect the tradition, and increased migration had brought people into the area who care nothing for *fadys* and who see an eager market available for the meat of the tortoise, not to mention the sale value of the creatures as pets. Even so, occasional collecting of the tortoises would not be devastating. But an industry has sprung up, and at the current rate that the tortoises are taken from the wild, even the largest, healthiest populations will be gone in a decade or two without intervention.

By emphasizing the historical ties between the local people and their tortoises, conservationists are attempting to restore respect for the *fady,* and to persuade local people that outsiders who come in and collect them are doing no service to the indigenous people. Teachers can do environmental outreach as well as anyone—they have all-day access to the young minds that will decide the future of the area and its animals. When local people consider themselves stewards of the land instead of exploiters, good things generally happen. If local people are stewards of the tortoises, the tortoises can be the flagship species of the entire ecosystem, as has been done in other parts of the world with more glamorous animals like pandas and tigers.

Local Heroes

Chimpanzees have Jane Goodall. Mountain gorillas had the late Dian Fossey. And the world's turtles and tortoises have Peter Pritchard. From a private research center in central Florida, Pritchard has conducted or inspired myriad conser-

vation projects the world over. His many books, from ency-
clopedic collections of chelonian information to personal
memoirs, have spurred younger generations of tortoise biologists for a generation. In the 1960s, Pritchard was a doctoral student working with the legendary sea turtle biologist
Archie Carr at the University of Florida at Gainesville. Pritchard's graduate work involved the study of sea turtle nesting
on the coast of Guyana in South America. In the process, he
learned more about nesting distributions for critically endangered sea turtle species in that region than anyone alive. He
also became immersed in conservation at an early point in his
career—doctorates are generally won or lost based on testing
grand theories, not trying to save the planet—when he en-
listed the help of local Arawak people in Guyana for his project to save and protect the turtles. He worked to turn the local
economy from one based on the value of sea turtle eggs to
poultry, and he hired locals to police the nesting beaches and
to be his research assistants. Changing an ancient culture in a
generation is a Sisyphean task, but Pritchard succeeded to the
point that Arawaks today regard protection of the local sea
turtles as a family tradition.

After his doctoral field research was over, Pritchard continued to conduct and supervise research on sea turtles. He
went to work for the State of Florida and also became involved
in research and conservation efforts on more far-flung members of the family, such as the giant tortoises of the Galápagos.
He became one of the main sleuths in the search for any remaining members of the Pinta Island giant tortoise subspecies (Chelonoidis nigra abingdoni). The one remaining representative of that race, the eponymous Lonesome George, had
been languishing in an enclosure at the Darwin Research Sta-

tion there in the faint hope that a female could be located traipsing the island. It was Pritchard who had been tipped off in 1972 to the existence of Lonesome George himself, in a casual mention of a large tortoise seen rambling around Pinta by a visiting biologist. That set off monumental efforts to locate Lonesome George, who was finally taken off Pinta the following year. Repeated expeditions since failed to turn up a single additional survivor. The last such trip, in 2002, was a systematic search of every inch of the island for not only living tortoises but their skeletons, droppings, and any other evidence of their recent existence. The search team turned up some remains of tortoises that may have perished only in the past several years, but no living tortoises.

Today, Peter Pritchard runs a privately funded research institute that he established in central Florida, not far from Disney World. He has amassed one of the world's largest collections of turtle and tortoise skeletal remains, representing nearly every known species. It's also a living museum, with giant tortoises rambling about. And Pritchard continues to be involved in conservation efforts around the world, most notably the effort to locate and reunite the few remaining specimens of the Yangtze giant softshell turtle in China, *Rafetus swinhoei*. These several-hundred-pound turtles, looking for all the world like gigantic manhole covers, once inhabited a wide area of southern China and northern Vietnam. After futile searches by Pritchard and others failed to turn up any wild Yangtze softshells, and after most of the remaining live *Rafetus* died in Chinese zoos, a final male and female were united in 2008. A first clutch of eggs failed to hatch, but hopes are now high that this eleventh-hour matchmaking will enable the species to have a future.

Peter Pritchard is the most visible of the global spokespersons for turtles and tortoises, but there are many others. Efforts to recover the giant softshell turtle have been directed in part by Gerald Kuchling, who has also devoted his recent career to finding the last remaining Burmese roofed river turtles *(Kachuga trivittata)* in Myanmar and directing the building of a breeding facility for them at the Yadanabon Zoo in Mandalay. Peter Paul van Dijk, a Dutch-born tortoise and turtle biologist based in Washington, D.C., is perhaps the foremost authority on the softshell and other river turtles of Asia. He has worked tirelessly to survey, document, and protect species in Asia and elsewhere. In China, the extinction epicenter for a great percentage of turtle biodiversity, Dr. Shi Haitao conducts conservation biology research on disappearing species and the role of turtle farms in saving or destroying species. He also trains the next generation of turtle biologists. He works with a number of western scientists conducting research on topics ranging from surveys of biodiversity to genetic investigations of the alleged new species coming out of China that may in fact be artfully produced hybrids.

Captive Caretakers

In addition to those conservationists who toil in mucky corners of the world, surveying or working to save tortoises in remote deserts, forests, and swamps, there is another army of conservationists working with tortoises in captivity who also play a vital role. As recently as the 1980s, nearly all reptiles and amphibians sold in the pet trade had been taken, legally or illegally, from the wild. This began to change when breakthroughs were made in creating an artificial hibernation environment for animals deprived of the crucial hormone-setting seasons in which they had evolved. Once breeding techniques were honed to produce reproduction-ready males and females, captive breeding was off to the races. What remained was discovering and endlessly tinkering with each species' requirements. The result of such labor was the ability to produce critically rare species outside their natural habitat.

Some private breeders have modest backyard facilities crammed with a representative of every rare tortoise species that they can find room for. Many opt for small species; others live in ways that most of us would not contemplate, devoting their daily lives to the care, feeding, and arduous moving of hundred-pound behemoths. Some specialize in one species only and have become world experts on its husbandry and breeding. All contribute to the storehouse of wisdom about how to care for tortoises. Those who emphasize breeding captive generations instead of removing breeding stock from the wild are genuinely moving a species painstakingly toward a future in captivity, if not in the wild.

The keeping and breeding of many of the more delicate species of tortoise is a craft bordering on an art. Temperature, humidity, diet, and a host of other factors spell the difference between a thriving and a languishing captive tortoise. Breeders become highly skilled veterinarians, parasitologists, and experts on animal behavior. Europeans mastered the art of creating a world inside a glass terrarium decades ago—even during Victorian times tiny rain forests in bell jars were all the rage. But the ability to make animals think they are in the wild to the point that they will reproduce is a recent innovation, one that followed centuries of putting specimens in zoos until they languished and died, only to be replaced by newly caught ones. Although my previous chapter about the pet trade should have made it clear that private collectors are sometimes the worst threat to the future of a critically rare species, they can also be saviors. These acquisitive folks wield a double-edged sword. The rapacious appetite some collectors have for depleting Nature of rare specimens barely keeps pace with their dedication to producing offspring from those prized possessions. Savvy, responsible breeders try to bring together animals that are not related to one another, which could make or break the future of the species. For some species for which captive breeding is the last hope, the obsessive time and energy that private breeders devote to their favorite species far outstrips the attention to one or a few species that most zoo curators can afford to give, with hundreds of animals under their care. The fate of the angonoka, for example, is entirely in the hands of captive breeders now. Other species may be added to that list in the coming years. In fact, the species that survive into the twenty-second century and those that don't may be decided by which species breed readily in captivity.

Unlike advocates for glamour animals like tigers and pandas, tortoise conservationists toil in relative obscurity. You don't see the icons of their devotion plastered on bumpers or in splashy magazine ads. Many fund their work with the help of very small private donations and even pay their own out-of-pocket expenses. Nearly all conservationists who work with large mammals are based at universities, zoos, or conservation organizations. In other words, saving species is their day job. But many chelonian conservationists are in the private sector, or are moonlighting in turtle and tortoise conservation efforts while earning their bread and butter in some other way. A network of zoo- and aquarium-based curators and keepers who breed the rarest species in captivity is assisted by an army of private citizens. Some tortoise conservation projects follow traditional notions of habitat preservation, captive breeding, and the like. But some put alternatives on the table. Biogeographer James Juvik of the University of Hawaii at Hilo and his colleagues Ross Kiester and Kenneth Nagy have proposed a head-start-and-release program for captive-bred desert tortoises in the California desert. Many release programs have existed before, but Juvik and colleagues' plan calls for a novel solution: asking the legions of private owners of desert tortoises across California to donate eggs their pets lay. They estimate a total of as many as a thousand eggs a year might be hatched, with offspring sent into a release program at selected sites in the desert. Such a massive program to give tortoises a head start has never been contemplated before, let alone tried. Currently, Californians can apply to adopt baby desert tortoises, which they raise in backyards with an agreement not to breed them. Plenty of breeding takes place nonetheless, and the idea of making use of these new genetic stocks for repatriation is intriguing.

It has also raised the ire of many desert tortoise conservationists, who consider the scheme ill conceived. They claim that it doesn't take into account the risk of contaminating wild tortoise populations with diseases spread from captivity, and such releases may mix genetically distinct tortoises. But Juvik and colleagues point to the very low likelihood that eggs, as opposed to live tortoises, would carry pathogens. They also argue that genetic mixing may even strengthen the tortoise stock and make it more adaptable to human-caused threats. A second problem is where exactly to head start all these tortoises—enormous enclosures in the desert where they learn to be wild? And then there is the vexing issue of where to release them. The new habitat must not have its own wild tortoise population, so diseases cannot be transmitted or territorial battles fought over existing burrows. But finding a suitable habitat from which all tortoises have already been extirpated is a Herculean task.

Adapting Strategy to Circumstance

Combating extinction risk requires this sort of creative thinking, even if only a fraction of such ideas turn out to be feasible. While land-management bureaucrats and academic biologists argue about the efficacy of someone's plan, the animals slide further toward the abyss. If the conservation problems of affluent countries, such as the California desert tortoise problem, seem intractable, consider how solutions may work in the developing world, where each conservationist is a much smaller voice in a much larger wilderness. Local people in developing countries have harvested tortoises for food for centuries. Until recently, consumption was fairly small scale and the human population density too low to be a mortal threat to the tortoises' continued survival. Local people may extol the virtues of the tortoise—its longevity, its toughness, its perceived wisdom of the ages. But that doesn't mean they won't eat them out of existence, or collect them and sell them to the pet trade until the last one is gone. Leopard tortoises, hardly one of the more endangered species, have been part of the rural economy of sub-Saharan Africa for millennia. Their meat is savored by everyone from the famed !Kung and Hadza hunter-gatherers to villagers the continent over. Their domed shells are used as water bowls and ladles, and in earlier times, as shields. But as human population density grew, depredations on the tortoise became too intense, too frequent to be sustainable.

And so the leopard tortoise, one of the commonest in many parts of Africa, has been put at risk of local extinction in some areas because of over-collecting. The value placed on the tortoise—respect for its longevity, a desire to preserve it as part of the culture—has been trumped by the need to eat. How can that pressure be reduced? If human population density continues to increase, only a change in local cultural values can restore a future to the leopard tortoise. The animal is prolific —it just needs a break to be able to cope with the intense pressure it is now experiencing.

In India, communities have long harvested river turtles for food. Changing the local culture from extractors to protectors is challenging, but some Indian herpetologists are succeeding. Shailendra Singh began a head-starting project with Indian river turtles, releasing thousands of turtles once they reached a size large enough to survive the rigors of the wild. That project has mushroomed into a model for community-based conservation action plans everywhere. The head-starting project provided employment to local people, who became ambassadors. They initiated workshops for local schoolchildren, employed and trained former turtle poachers to be turtle surveyors and protectors, and worked with fishermen to adopt more turtle-friendly catch techniques. These actions have all translated into a successful conservation plan contributing to the future of all local fauna.

In Egypt, the tiny Egyptian tortoise has been whittled to near-extinction by massive collecting for the pet trade. Now some of the same people who would have been out searching for the tortoises to send to dealers are involved in their conservation. Even though tortoises as rare as the Egyptian bring hundreds of dollars on the global pet market, they are worth only a tiny fraction of that amount to the local people who collect them. So offering those same folks a replacement salary to work on behalf of the tortoises—whether it is surveying and taking a census, or rearing hatchlings in captivity—can provide adequate incentive for them not to collect them. The budget for such a project need not be large because the locals were not getting rich off the tortoises anyway. Only the middlemen get rich, and the money is not passed on to the local people.

Involving the community is one very sound conservation strategy. But tortoises are only a small part of any ecosystem; when the tortoises are facing extinction, it is very likely that many other species are, too. So combining a tortoise and turtle conservation effort with larger conservation goals and strategies makes a great deal of sense. When you protect an ecosystem, you create an umbrella under which many animals and plants are saved. That can be combined with a public education and outreach program such as those described in this chapter. Finally, in a few cases, tortoises can be the focus of ecotourism.

Big animal ecotourism works very well in the right setting. In East Africa, ecotourists flock to a few isolated islands of tropical forest high in the mountains of the Albertine Rift to see the world's remaining mountain gorillas. They pay four hundred dollars or more for one hour of gorilla watching, which requires them to rise at dawn and trek with a guide over ridge and valley until they find a group of gorillas. The apes have been slowly habituated to being approached by humans, and for a magical sixty minutes a small band of high-rolling humans gets to watch the silverback loll, the mothers dote, and the infants play. The forest around them is replete with thousands of other animals and plants—the region is one of the most biologically unique and diverse on Earth—but folks come for the gorillas. A small percentage of the fees they pay are dedicated to local villages for the construction of clinics, dispensaries, and the like. It is, in theory, a win-win situation in conservation terms. Some experts worry, however, about unwarranted stress on the gorillas, and certainly there is a heightened risk of contagious diseases being passed from human to gorilla, to the detriment of the animals.

In other parts of Africa wildlife watching has been big business for decades; the safari crowd in fact sometimes outnumbers the lions at a kill. And glamour animal ecotourism is big not only in Africa. The millions who flock to Yellowstone National Park every year come for a glance at a grizzly bear or a wolf, not just for the mountain air. On a few barrier islands off the mid-Atlantic coast, wild ponies attract tourists. So wildlife of all sorts can be big business, and where there is money to be made and the system is carefully managed, tourism can be a savior.

Tortoise ecotourism works well in two widely separated parts of the world. The Galápagos are visited by tens of thousands of people every year for far more than just their gigantic tortoises. But giant tortoises remain the most enduring icon of the islands. They are the centerpiece of tourism on Santa Cruz Island, where visitors can hike up into the hills to see them in their habitat or hang out with them at the captive breeding center in Puerta Ayora. It's safe to say that without their tortoises, Ecuadorians would reap far less economic windfall from the Galápagos Islands. The situation is different in the Indian Ocean. The Seychelles and Mascarene Islands are already vacation spots, well known to Europeans for their beaches and snorkeling spots. Tortoise ecotourism is a nice complement to tourism there, providing tourists with a day trip to a captive facility or perhaps to one of the more far-flung islands repopulated with tortoises. As we saw earlier, the island of Rodrigues, which had not had a natural population of giant tortoises in two centuries, now has a small but growing population of introduced Aldabras, and tour buses arrive all day long to visit them.

But Galápagos and Aldabra tortoises are two of the most impressive animals on Earth. Could tortoise ecotourism work with less glamorous varieties? It likely would work if the species in question were one of the giants. There are many habitats in which tropical giant tortoises could thrive, such as Hawaii, southern California, Florida, the Caribbean, and the Mediterranean, to name a few. Some of these—Florida and the Caribbean—hosted their own giant tortoise species in the not-so-distant past. So how about introducing endangered large tortoises to areas where they are not native? The idea of intentionally creating nonnative populations of animals, even rare ones, makes many conservationists blanch, but the idea has merit. The benefit is the creation of a new population safe from rising ocean levels or tsunamis that might threaten to wipe out the one or few remaining natural populations. The risk is always that of unintended consequences. Unlike other relocated species that have become pests without natural predators—think rabbits in Australia or pythons in the Florida Everglades—tortoises aren't going anywhere. They can't run away and they don't hide very well. If introducing them turned out to be a bad idea for their new habitat, collecting them again would not be difficult. Giant tortoises are demonstrably good at drawing tourists, so why wouldn't a resort hotel in the Caribbean or Florida want a few roaming around their grounds?

If this logic could be extended to some of the smaller but equally eye-catching species, it could be a boon for ex-situ conservation. Imagine having eighty-pound Asian forest tortoises living in humid areas of the Caribbean or gorgeous radiated tortoises thriving in the scrubby arid lands outside Los Angeles. There would be biological and bureaucratic hurdles to surmount, but there are certainly cases in which it would work. It would require taking tortoise conservation out of the hands of professionals and accepting the aid of resort owners, private landowners, and others whom the conservation community has long distrusted with any endangered animal issue. But as a potential aid in developing assurance colonies of rare species all over the world in appropriate habitats, if not traditional geographic strongholds, global translocation might save many species from extinction.

ACHILLES AND THE TORTOISE

Lewis Carroll wrote of a bet that defied all logic. A tortoise challenged Achilles—the heroic figure of the Iliad in Greek mythology—to a race. The tortoise, feeling confident, bet that he could beat Achilles if he were given a modest head start. Achilles found this proposition hilarious, and told the little tortoise to take just as big a head start as he needed. The tortoise asked for a mere ten meters on a one-hundred-meter track. Achilles accepted. Then the tortoise pointed out that in the time it took Achilles to catch up, the tortoise would have advanced one extra meter toward the finish line. By the time Achilles made up that extra distance, the tortoise would have gained a fraction of another meter. By the time Achilles made up that fraction, the tortoise would have moved another very small fraction farther. After some contemplation, Achilles conceded the race, because overtaking the tortoise began to seem to defy the laws of motion.

This paradox does not truly defy the laws of motion, of course, it just gives that appearance. In the real world, humans have overtaken the tortoise in a stunningly short time, a fraction of an instant in evolutionary time. For the tortoise, we are the cataclysm, the End Times. Some species will survive the present apocalypse; many will not. Many more will continue to exist only as captive species; pretty birds in gilded cages, as it were. There are popular pet species that are still relatively abundant in the wild today that will in the future cost a month's salary, if they are available at all. Many of the tortoises with low reproductive rates or low fertility in captivity will, once the wild populations are virtually exterminated, hang on only in zoos and in the hands of wealthy private collectors. They will no longer be a species in the evolutionary sense. They will just be a scattered gene pool, a few protected, priceless animals locked up in cages.

In the developed world, there will be valiant rescue efforts just as there are today, and some species, like the American desert tortoise, will endure simply because enough people care that it not slip into extinction. In the developing world, however, wholesale, rapid habitat destruction combined with pillaging the tortoise populations themselves for the pet trade and for food will depopulate the vast majority of the homes of dozens of species. Most of the damage will be done before anyone can do much to prevent it. In Asia, the Chinese extinction juggernaut will drive nearly all of the current species living in Asia into extinction.

It doesn't have to be this way. In an ideal world of the future, even developing countries would have cultivated an ethos of enjoying rather than exploiting the natural world in which the tortoises live. Large tracts of rain forest would have been set aside, harboring not only tortoises but myriad other animals large and small. Instead of logging the forests as fast as possible, they could be kept intact as carbon absorbers. Governments could build infrastructure to promote tourism, and ecotourists would flock to the national park to camp, hike, birdwatch, and enjoy the great outdoors. Instead of taking home any living thing they can find, the ecotourists would take only photographs. The only risk to a tortoise lumbering across its ancient world would be the occasional road crossing or the occasional camping family who decided to pick up a tortoise as a pet. The money earned from carbon trading or from tourism would provide more than enough incentive for the government to continue to invest in the forest—not as a timber resource, but as an ecosystem services provider and a tourism and scientific research resource. The tortoises would be saved in perpetuity as part of the natural heritage of the nation.

In fact, this is exactly the situation in Kaeng Krachan National Park in southwestern Thailand. The park is one of the largest in Southeast Asia and is a nearly intact ecosystem. It features one of the largest tiger populations in any nature reserve anywhere, plus elephants and nearly every other large Asian mammal species. It also is a stronghold of the Asian forest tortoise (*Manouria emys*), the largest Asian tortoise and a denizen of mountainous wet forests from eastern India through Indonesia. The tortoise lives at high densities in the ravines, streams, and swamps of the watershed just below the highest point in the park, where our study has been situated since 2005. Pratyaporn Wanchai, a Thai graduate student from Chulalongkorn University in Bangkok, has, in collaboration with Dr. Kumthorn Thirakupt and myself, been conducting the first detailed study of this species. This largest Asian tortoise has been barely studied, even though its prime habitat is an easy three-hour drive from one of the world's largest cities.

The Asian forest tortoise is to Thais what the desert tor-

toise is to southern Californians: their adorable shelled critter. They are found crossing roads, picked up, and kept as pets.
Unlike the desert tortoise, they are also eaten by rural people.
Kaeng Krachan is large enough and well-patrolled enough
that in the heart of the park, safely tucked under the hilltop
housing the ranger staff, the tortoises are quite safe. We see
them living as they must have for millennia before humans
encroached on their rapidly vanishing habitat. Their ability to negotiate their fifty-pound bulk up and over mountain
slopes so steep that I need to grip rocks and branches with my
arms just to haul myself up is legendary. Their Thai name, *tao
hoq,* means "six-legged tortoise," a reference to the enormous
thornlike spurs of scaly flesh that jut out from the area between their actual rear legs and tail. More likely an antipreda-
tor defense, the spurs might lead anyone seeing the death-
defying scaling of cliffs and hills these tortoises can achieve
to believe that the tortoises possess extra mountain-climbing
gear.

Middle-class Thais have embraced ecotourism, and weekends during the cool winter season find campgrounds over-
flowing with tents, each with a family grilling food, playing
with their kids, and generally enjoying the great outdoors.
While many other Asian countries lack such a homegrown
tourist industry, and locals tend to see the forest as a for-profit
resource, sunrise in Kaeng Krachan finds Thai ecotourists
from Bangkok meditating, sitting at easels painting watercolors of the landscape, and ticking off each new bird species
they can see. It is the western model of conservation—raising
the public's consciousness—and it has begun to take hold in
Thailand. Thais tend to be a law-abiding people; they respect
the land and see the preservation of natural habitat as a long-

term investment. This could not be more at odds with the view of some of their close neighbors in Asia. Within easy walking distance of one of Kaeng Krachan's campgrounds, Asian forest tortoises lumber through the forest, leading lives in relative peace and tranquility, relatively safe from marauding loggers, poachers, and the like.

Not all is perfect, of course. Kaeng Krachan is only a few years removed from intense poaching pressure, and illegal hunting no doubt continues along the boundaries of the park, just as it does in Yellowstone and many other sanctuaries in the western world. Thailand is still a developing country by most measures, though certainly not in the same category as much of Southeast Asia or Africa. It has an economic infrastructure and literacy rate strong enough to allow environmental values to flourish, and a culture that places a high value on quality of life, influenced by its Buddhist sensibilities. It is governed by a royal family to whom profound respect is given. If the king tells you to respect Nature, you'll respect Nature. So Thais may be exceptional, but they are by no means unique. And it is a recipe that can work elsewhere.

The scenario varies country by country according to its culture, its level of affluence or poverty, the competing values of forest cutting or poaching, and the importance placed on preserving the forest and its wildlife for the sake of future generations. The Thai ethic is not going to take hold in the Democratic Republic of Congo anytime in the foreseeable future. Poverty there is desperate, political stability and infrastructure are almost nonexistent, and the economic incentive of ecotourism is only a fleeting dream interrupted continually by civil war and ensuing chaos. In many other parts of the African continent, ecotourism is a highly successful tool for both ensuring conservation management and enriching local people and the government that sponsors it. But the tourists are almost entirely foreigners, and to say that a conservation ethic has taken hold would be an overstatement. Take away the tourism, even temporarily, and the forest and its animals immediately become exploitable commodities again. A conservation program that works because the people of the area want it to succeed is far more promising than one that runs on the backs of armed park rangers barring local people from the forest with threat of arrest.

Respect for tortoises is one thing, but getting to the point at which local people have the luxury of protecting them, or at least declining to exploit them, is another. The competing interests of land use and food for hungry villagers combined with the money to be earned by selling the animals into the global pet trade has brought most of the world's tortoises to death's door. The slide to extinction is reversible, but not for long. Words mean a great deal only until the word has gotten out that a terrible problem exists that can still be fixed. Once the problem has festered so long that solutions are no longer on the table, the words ring hollow. But words can help bring action, and that action must not be too late to save these most cherished and unique of Earth's creatures from becoming faded photos in children's picture books.

Like the dodo, tortoises are priceless works of art. People of the next millennium will look at the few remaining examples and be saddened the people of our century failed to value them enough to save them. Unless we act now, the last remnants of a once-great flourishing of uniquely strange and wonderful creatures will disappear from Earth in short order. They and we deserve a far better fate.

Appendixes

Further Reading

Acknowledgments

Index

Extremes of the Tortoise World

Species	Maximum length	Maximum weight
speckled padloper	4 inches	< 1 pound
Nama padloper	4.5 inches	< 1 pound
spider tortoise	6 inches	1 pound
Egyptian tortoise	5 inches	1 pound
Aldabra giant tortoise	> 3 feet	700 pounds?*
Galápagos tortoise	> 3 feet	900 pounds?**
leopard tortoise	2.5 feet	100 pounds
Asian forest tortoise	2 feet	80 pounds
spurred tortoise	3 feet	> 200 pounds

*An adult male is more typically 350–400 pounds.

**An adult male of the largest populations is more typically 400 pounds.

Tortoises on the Brink of Extinction

These tortoises (listed with their geographic ranges) are on the brink of extinction in the wild. The main forces that threaten them are recorded here.

angonoka or ploughshare tortoise (western Madagascar): pet trade, habitat loss

Burmese star tortoise (Myanmar): pet trade, habitat loss

Egyptian tortoise (coastal North Africa, Middle East): pet trade, habitat loss

flat-tailed spider tortoise (western Madagascar): habitat loss, pet trade

geometric tortoise (South Africa): habitat loss

impressed tortoise (Southeast Asia): food markets

Nama padloper (Namibia): tiny range

pancake tortoise (East Africa): pet trade, habitat loss

radiated tortoise (southern Madagascar): pet trade, food markets

San Salvador giant tortoise (Galápagos Islands): nineteenth-century slaughter

Santa Cruz Island giant tortoise (Galápagos Islands): nineteenth-century slaughter

southern speckled padloper (South Africa): habitat loss

spider tortoise (southwestern Madagascar): pet trade, habitat loss, human consumption

Volcán Sierra Negra giant tortoise (Galápagos Islands): nineteenth-century slaughter

Tortoise Species

This list includes currently recognized tortoise species and subspecies and their ranges, including recently extinct tortoise species. Adapted from Fritz and Havas (2007).

Astrochelys
Astrochelys radiata: Madagascar
Astrochelys yniphora: Madagascar

Cylindraspis
Cylindraspis indica: Réunion (extinct 1840)
Cylindraspis inepta: Mauritius (extinct 1735)
Cylindraspis triserrata: Mauritius (extinct 1735)
Cylindraspis peltastes: Rodrigues (extinct 1795)
Cylindraspis vosmaeri: Rodrigues (extinct 1795)

Chelonoidis
Chelonoidis carbonaria: northern South America
Chelonoidis chilensis: Argentina
Chelonoidis denticulata: northwestern South America
Chelonoidis nigra abingdoni: Pinta Island (Lonesome George)
Chelonoidis nigra becki: Isabela Island (Volcán Wolf)
Chelonoidis nigra chathamensis: San Cristóbal Island (extinct 1933)
Chelonoidis nigra darwini: Santiago Island

Chelonoidis nigra elephantopus: Isabela Island (Volcán Cerro
    Azul)
Chelonoidis nigra ephippium: Pinzón Island *(duncanensis)*
Chelonoidis nigra guntheri: Isabela Island (Volcán Sierra Negra)
Chelonoidis nigra hoodensis: Española Island
Chelonoidis nigra microphyes: Isabela Island (Volcán Darwin)
Chelonoidis nigra nigra: Floreana Island (extinct 1850)
Chelonoidis nigra porteri: Santa Cruz Island
Chelonoidis nigra phantastica: Fernandina Island (extinct)
Chelonoidis nigra vandenburghi: Isabela Island (Volcán Alcedo)

Chersina
Chersina angulata: South Africa

Dipsochelys
Dipsochelys dussumieri: Aldabra

Geochelone
Geochelone elegans: India, Sri Lanka
Geochelone platynota: Myanmar
Geochelone sulcata: Africa

Gopherus
Gopherus agassizii: southwestern United States
Gopherus berlandieri: Texas, northern Mexico
Gopherus flavomarginatus: north central Mexico
Gopherus polyphemus: southeastern United States

Homopus
Homopus areolata: South Africa, Cape
Homopus boulengeri: South Africa, Karoo
Homopus femoralis: South Africa, Cape
Homopus signatus cafer: northwestern South Africa
Homopus signatus signatus: Namibia? South Africa
Homopus solus: Namibia

Indotestudo

Indotestudo elongata: Southeast Asia
Indotestudo forstenii: Sulawesi
Indotestudo travancorica: southwestern India

Kinixys

Kinixys belliana belliana: West, East Africa
Kinixys belliana domerguei: Madagascar (introduced)
Kinixys belliana mertensi: Democratic Republic of the Congo
Kinixys belliana nogueyi: Senegal, Cameroon, Central African
   Republic
Kinixys belliana zombensis: Tanzania, Zululand
Kinixys erosa: Gambia, Uganda
Kinixys homeana: Liberia, Democratic Republic of the Congo
Kinixys lobatsiana: Botswana
Kinixys natalensis: southeastern Africa
Kinixys spekii: Angola, Kenya

Malacochersus

Malacochersus tornieri: southern Kenya, Tanzania, Zambia

Manouria

Manouria emys emys: Southeast Asia
Manouria emys phayrei: mainland Southeast Asia
Manouria impressa: Myanmar, Malaysia

Psammobates

Psammobates geometricus: southwestern Cape, South Africa
Psammobates oculifer: South Africa, Namibia, Botswana
Psammobates tentorius tentorius: southern, eastern Karoo
Psammobates tentorius trimeni: western Cape, South Africa
Psammobates tentorius verroxii: northern Cape, South Africa,
   Namibia

Pyxis

Pyxis arachnoides arachnoides: southwestern Madagascar

Pyxis arachnoides brygooi: southwestern Madagascar

Pyxis arachnoides oblonga: southeastern Madagascar

Pyxis planicauda: southwestern Madagascar

Stigmochelys

Stigmochelys pardalis babcocki: East Africa

Stigmochelys pardalis pardalis: South Africa

Testudo

Testudo (Agrionemys) horsfieldii horsfieldii: Afghanistan, Iran, China

Testudo (Agrionemys) horsfieldii kazakhstanica: Kazakhstan

Testudo (Agrionemys) horsfieldii rustamovi: Turkmenistan

Testudo (Testudo) graeca anamurensis: Turkey

Testudo (Testudo) graeca antakyensis: Turkey

Testudo (Testudo) graeca armeniaca: Armenia, Iran

Testudo (Testudo) graeca cyrenaica: Libya

Testudo (Testudo) graeca floweri: Israel, Lebanon

Testudo (Testudo) graeca graeca: northeastern Morocco

Testudo (Testudo) graeca ibera: Azerbaijan, Iran

Testudo (Testudo) graeca lamberti: northern Morocco

Testudo (Testudo) graeca marokkensis: north central Morocco

Testudo (Testudo) graeca nabeulensis: Tunisia

Testudo (Testudo) graeca nikolskii: Russia

Testudo (Testudo) graeca pallasi: Russia

Testudo (Testudo) graeca perses: Turkey, Iran

Testudo (Testudo) graeca souissensis: southwestern Morocco

Testudo (Testudo) graeca terrestris: Turkey, Syria

Testudo (Testudo) graeca zarudnyi: Iran

Testudo (Testudo) hermanni hermanni: France, Spain

Testudo (Testudo) hermanni boettgeri: Italy, Turkey

Testudo (Testudo) kleinmanni: Egypt, Libya, Israel

Testudo (Testudo) marginata: Greece, Albania

FURTHER READING

1. What Exactly Are Tortoises and Turtles?

Foster, W., ed. 1905. *The journal of John Jourdain 1608–1617, describing his experience in Arabia, India, and the Malay Archipelago.* Cambridge: Hakluyt Society.

Fritz, U., and P. Havas. 2007. Checklist of the chelonians of the world. *Vertebrate Zoology* 57: 149–368.

Li, C., X. Wu, O. Rieppel, L. Wang, and L. Zhao. 2008. An ancestral turtle from the Late Triassic of southwestern China. *Nature* 456: 487–501.

Reisz, R., and J. J. Head. 2008. Turtle origins out to sea. *Nature* 456: 450–451.

2. Live Long and Prosper

Belzer, B. 1999. Survival among translocated eastern box turtles. *Box Turtle Research and Conservation Newsletter* 8: 7–8.

Congdon, J. D., and J. W. Gibbons. 1987. Morphological constraint on egg size: A challenge to optimal egg size theory. *Proceedings of the National Academy of Science* 16: 4145–4147.

Congdon, J. D., R. D. Nagle, O. M. Kinney, M. Osentoski, H. Avery, R. C. van Loben Sels, and D. W. Tinkle. 2000. Nesting ecology and embryo mortality: Implications for the demography of Blanding's turtles (*Emydoidea blandingii*). *Chelonian Conservation and Biology* 3: 569–579.

Congdon, J. D., R. D. Nagle, O. M. Kinney, and R. C. van Loben Sels. 2001.

Hypotheses of aging in a long-lived vertebrate, Blanding's turtle (*Emydoidea blandingii*). *Experimental Gerontology* 36: 813–827.

Germano, D. J. 1992. Longevity and age-size relationships of populations of desert tortoises. *Copeia* 2: 367–374.

Gibbons, J. W., and R. D. Semlitsch. 1982. Survivorship and longevity of a long-lived vertebrate species: How long do turtles live? *Journal of Animal Ecology* 51: 523–527.

Heppell, S. S., L. B. Crowder, and D. T. Crouse. 1996. Models to evaluate headstarting as a management tool for long-lived turtles. *Ecological Applications* 6: 556–565.

Lanner, R. M., and K. F. Connor. 2001. Do bristlecone pines senesce? *Experimental Gerontology* 36: 675–685.

O'Brien, S., R. Bourou, and T. Hafany. 2005. Hatch size, somatic growth rate and size-dependent survival in the endangered ploughshare tortoise. *Biological Conservation* 126: 141–145.

3. No Respect for the Ancient Lands

Adams, J. S., and T. O. McShane. 1992. *The myth of wild Africa.* Berkeley: University of California Press.

Aponte, C., G. R. Barreto, and J. Terborgh. 2003. Consequences of habitat fragmentation on age structure and life history in a tortoise population. *Biotropica* 35: 550–555.

Baard, E. H. W. 1993. Distribution and status of the geometric tortoise *Psammobates geometricus* in South Africa. *Biological Conservation* 63: 235–239.

Boarman, W. I., and M. Sazaki. 2006. A highway's road-effect zone for desert tortoises (*Gopherus agassizii*). *Journal of Arid Environments* 1: 94–101.

Boycott, R. C., and O. Bourquin. 2000. *The southern African tortoise book.* KwaZulu-Natal, South Africa: O. Bourquin.

Ferraz, G., G. J. Russell, P. C. Stouffer, R. O. Bierregaard, Jr., S. L. Pimm, and T. E. Lovejoy. 2003. Rates of species loss from Amazonian forest fragments. *Proceedings of the National Academy of Sciences* 100(24): 14069–14073.

McCoy, E. D., and H. R. Mushinsky. 1999. Habitat fragmentation and the abundances of vertebrates in the Florida scrub. *Ecology* 80(8): 2526–2538.

Tuberville, T. D., T. M. Norton, B. D. Todd, and J. S. Spratt. 2008. Long-term apparent survival of translocated gopher tortoises: A comparison of newly released and previously established animals. *Biological Conservation* 141: 2690–2697.

4. Eating Tortoises

Brennessel, B. 2006. *Diamonds in the marsh.* Lebanon, N.H.: University Press of New England.

Gong, S., J. Wang, H. Shi, R. Song, and R. Xu. 2006. Illegal trade and conservation requirements of freshwater turtles in Namao, Hainan Province, China. *Oryx* 40: 331–336.

Nussbaum, R. A., and C. J. Raxworthy. 2000. Commentary on conservation of *"Sokatra,"* the radiated tortoise (*Geochelone radiata*) of Madagascar. *Amphibian and Reptile Conservation* 2(1): 6–14.

Parham, J. F., W. B. Simison, K. H. Kozak, C. R. Feldman, and H. Shi. 2001. New Chinese turtles: Endangered or invalid? A reassessment of two species using mitochondrial DNA, allozyme electrophoresis and known-locality specimens. *Animal Conservation* 4: 357–367.

Schafer, E. H. 1962. Eating turtles in ancient China. *Journal of the American Oriental Society* 82(1): 72–73.

Sharma, D. S. K., and O. B. Tisen. 2000. Freshwater turtle and tortoise utilization and conservation status in Malaysia. *Chelonian Research Monographs* 2: 120–128.

Shepherd, C. R. 2000. Export of live freshwater turtles and tortoises from North Sumatra and Riau, Indonesia: A case study. *Chelonian Research Monographs* 2: 112–119.

Shi, H., J. F. Parham, F. Zhiyong, H. Meiling, and Y. Feng. 2008. Evidence for the massive scale of turtle farming in China. *Oryx* 42: 147–150.

Shi, H., J. F. Parham, M. Lau, and C. Tien-His. 2007. Farming endangered turtles to extinction in China. *Conservation Biology* 21: 5–6.

Zhou, Z., and Z. Jiang. 2008. Characteristics and risk assessment of international trade in tortoises and freshwater turtles in China. *Chelonian Conservation and Biology* 7: 28–36.

5. "Such Huge Defourmed Creatures"

Adams, D. 1990. *Last chance to see.* New York: Ballantine Books.

Austin, J. J., and E. N. Arnold. 2001. Ancient mitochondrial DNA and morphology elucidate an extinct island radiation of Indian Ocean tortoises (*Cylindraspis*). *Proceedings of the Royal Society of London, Series B* 268: 2515–2523.

Austin, J. J., E. N. Arnold, and R. Bour. 2003. Was there a second adaptive radiation of giant tortoises in the Indian Ocean? Using mitochondrial DNA to investigate speciation and biogeography of *Aldabrachelys* (Reptilia, Testudinidae). *Molecular Ecology* 12: 1415–1424.

Caccone, A., G. Gentile, J. P. Gibbs, H. L. Snell, J. Betts, and J. R. Powell. 2002. Phylogeography and history of giant Galapagos tortoises. *Evolution* 56: 2052–2066.

Gerlach, J. 2004. *Giant tortoises of the Indian Ocean.* Frankfurt: Chimaira.

Gerlach, J., and K. L. Canning. 1998. Taxonomy of Indian Ocean giant tortoises (*Dipsochelys*). *Chelonian Conservation and Biology* 3: 3–19.

Gerlach, J., C. Muir, and M. Richmond. 2006. The first substantiated case of trans-oceanic tortoise dispersal. *Journal of Natural History* 40: 2403–2408.

Hansen, D. M., C. N. Kaiser, et al. 2008. Seed dispersal and establishment of endangered plants on oceanic islands: The Janzen-Connell model, and the use of ecological analogues. *PLoS ONE* 3(5): e2111.

Karanth, K. P., E. Palkovacs, J. Gerlach, S. Glabermann, J. P. Hume, A. Caccone, and A. D. Yoder. 2005. Native Seychelles tortoises or Aldabran imports? The importance of radiocarbon dating for ancient DNA studies. *Amphibia-Reptilia* 26: 116–121.

Nichols, H. 2006. *Lonesome George: The life and loves of a conservation icon.* New York: Macmillan.

Palkovacs, E. P., M. Marschner, C. Ciofi, J. Gerlach, and A. Caccone. 2003. Are the native giant tortoises from the Seychelles really extinct? A genetic perspective based on mtDNA and microsatellite data. *Molecular Ecology* 12: 1403.

Parham, J. F. 2008. Rediscovery of an "extinct" Galapagos tortoise. *Proceedings of the National Academy of Sciences* 105: 15227–15228.

Poulakakis, N., S. Glaberman, M. Russello, L. B. Beheregaray, C. Ciofi, J. R. Powell, and A. Caccone. 2008. Historical DNA analysis reveals living descendants of an extinct species of Galapagos tortoise. *Proceedings of the National Academy of Sciences* 105: 15464–15469.

Russello, M. A., L. B. Beheregaray, J. P. Gibbs, T. Fritts, N. Havill, J. R. Powell, and A. Caccone. 2007. Lonesome George is not alone among Galápagos tortoises. *Current Biology* 17: R317–318.

Stoddart, D. R., and J. F. Peake. 1979. Historical records of Indian Ocean giant tortoise populations. *Philosophical Transactions of the Royal Society of London. Series B* 286: 147–161.

6. Beloved Captives

Chatfield, J. 1986. Selborne and the tortoise connection. *Testudo* 2(4): 15–35.

Fitzgerald, S. 1989. *International wildlife trade: Whose business is it?* Washington, D.C.: World Wildlife Fund.

Klemens, M. W., and D. Moll. 1995. An assessment of the effects of commercial exploitation on the pancake tortoise, *Malacochersus tornieri,* in Tanzania. *Chelonian Conservation and Biology* 1: 197–206.

Leuteritz, T. E. J., T. Lamb, and J. C. Limberaza. 2005. Distribution, status and conservation of radiated tortoises (*Geochelone radiata*) in Madagascar. *Biological Conservation* 124: 451–461.

O'Brien, S., E. R. Emahalala, V. Beard, R. M. Rakotondrainy, A. Reid, V. Raharisoa, and T. Coulson. 2003. Decline of the Madagascar radiated tortoise (*Geochelone radiata*) due to overexploitation. *Oryx* 37: 338–343.

Warner, S. T. 1946. *Portrait of a tortoise.* Repr., London: Virago, 1981.

7. Are There Solutions?

Juvik, J. O., D. E. Meier, and S. McKeown. 1991. Captive husbandry and conservation of the Madagascar angulated tortoise, *Geochelone yniphora. Proceedings of the First International Symposium on Turtles and Tortoises: Conservation and Captive Husbandry,* 127–137.

Kabigumila, J. 1998. Community attitudes to tortoises (*Geochelone pardalis babcocki*) and their conservation in northern Tanzania. *African Study Monographs* 19(4): 201–216.

ACKNOWLEDGMENTS

When I began the research on tortoises that led to this book, I was worried that the community of scientists already working in the field might not be open to a latecomer poking around asking for information and advice. Academic territoriality runs as deep as human nature allows. But a large number of people gave of their advice and their time in the most generous ways possible, which allowed me to rapidly turn ideas for research projects in Southeast Asia on two obscure species (the Asian forest tortoise, *Manouria emys,* and the impressed tortoise, *Manouria impressa*) into reality. After a few years working with new colleagues and students, I got the idea of writing this book in the hope that an enlightened public could become more aware of and more engaged in the extinction crisis facing tortoises. Again, the specter of academic territoriality reared its head. Who was I to write about work that others have been sweating over for decades? But again, I received nothing but encouragement and cheerfully constructive advice from a wide range of professional biologists, conservationists, zoo curators and keepers, and private tortoise breeders.

In Asia, I am grateful for the help of my colleague Professor

Kumthorn Thirakupt of Chulalongkorn University in Bangkok, a leading authority on turtles and tortoises of the region. Varanya Aranyavalai of the same department also provided assistance, and one of Kumthorn's and my students, Pratyaporn Wanchai, carried out research for his master's degree on Asian forest tortoises in southwestern Thailand. For their assistance and advice in educating me about setting up a radio-telemetry project in the middle of a Thai rain forest, I thank the late John Behler, Bill Belzer, Craig Guyer, Doug Hendrie, Klaus Hoybe-Mortensen, Steve Platt, Bryan Stuart, and Peter Paul van Dijk. For their hospitality in Myanmar, I thank Khin Myo Myo, Win Ko Ko, and the staff at the headquarters of the Wildlife Conservation Society in Yangon; Professor Daw Tin Nwe of Yangon University; and Dr. Tint Lwin of the Yangon Zoo, as well as my traveling companion there, Alan Mootnick. In Cambodia I am very grateful to David Emmett and his staff for much-needed assistance in surveying for and setting up a field project on the impressed tortoise, and to Chey Koulang for his excellent field research on this beautiful and bizarre species. I also thank William Espenshade and Tim McCormack for help in setting up the project with Chey Koulang in Cambodia.

For general discussions and advice about the natural history of the Asian forest tortoise, thanks go to Eric Akaba, Vic Morgan, Jonathan Murray, and Chuck Schaeffer. For the same sort of crucial information on the impressed tortoise in captivity I thank Janet Eldredge, Eric Goode, Ralph Hoekstra, and Dwight Lawson.

During a sabbatical trip to the Mascarene Islands I received much hospitality and advice from the conservationists engaged in trying to bring the ecosystems and tortoise fauna of Mauritius and Rodrigues back from the dead. On Mauritius I thank Christine Griffiths for a day at the habitat restoration project on Île aux Aigrettes and Vikash Tatayah for a visit to the Mauritian Wildlife Foundation's captive breeding center at Black River Gorges. Owen Griffiths gave me his time and a wealth of information at the Aldabra tortoise breeding facility at La Vanille. Aurele Andre and his staff at the François Leguat Giant Tortoise Reserve, the brainchild of Owen Griffiths, had me sitting among a hundred giant tortoises when they weren't lowering me into limestone caves in search of rare bones.

For taking the time and trouble to review and critique the book manuscript I owe many thanks to Drs. Dennis Hansen, James Juvik, and Peter Paul van Dijk. Further information came from correspondence and discussions with Drs. James Juvik, Peter Pritchard, Anders Rhodin, and Shi Haitao. Photos of angonoka and spider tortoises in Madagascar were kindly provided by Peter Paul van Dijk. Thanks once again to Michael Fisher at Harvard University Press for his repeated enthusiasm for my work and his faith in the value of books of this genre, and to Russell Galen of Scovil, Galen, Ghosh literary agency.

As always, I owe a debt of gratitude to my ever-patient and supportive family: Gaelen, Marika, Adam, and especially my wife, Erin.

Proceeds from sales of this book are being used to support tortoise conservation projects in Asia and elsewhere.

INDEX

Adams, Douglas, 121
Adams, Jonathan, 71
Aesop, 9, 54
Africa, 3, 13, 21–26, 28–29, 31, 34, 44, 67, 69, 78, 85–86, 88, 101, 137, 149–153, 160, 175–178, 185–186
Age, determination of, 39
Agriculture, 70, 85, 87, 163
Aldabra tortoises, 21, 25, 36–37, 48, 53–54, 110, 113–114, 116–117, 119, 129, 135–137, 143, 161, 179
Amazon, 23–24, 68, 76
Amphibians, 156
Angonoka. See Ploughshare tortoises
Ankarafantsika National Park, 164
Anthodon, 18
Arawaks, 169
Argentina, 109
Arizona, 79
Asia, 1–3, 19, 21, 23–24, 26, 29, 31, 55, 65–66, 68, 70, 88–89, 91, 93–96, 98, 101–107, 139–142, 144, 148–149, 170–171, 182–185

Asian forest tortoises, 19–20, 24, 33, 44, 55, 91–92, 102, 180, 183–184
Astrochelys, 22. See also Ploughshare tortoises; Radiated tortoises
Austin, Jeremy, 115
Australia, 20, 79

Bacteria, 81–82
Bangkok, Thailand, 139–141
Bangladesh, 88–89, 101
Beagle, H.M.S., 35–36, 128
Bell's hingeback tortoises, 24
Biodiversity, 3, 22, 64–66, 75–76, 107, 148, 171
Blanding's turtles, 49–52
Body temperature, 13, 16–17
Bolson's tortoises, 22
Box turtles, 11–12, 16, 42, 55, 103, 144–146, 159
Brazil, 75
Bristlecone pines, 39
Brumation, 17, 25

Buddhism, 105, 144, 185

Burma. *See* Myanmar

Burmese star tortoises, 23, 146

"Bushmeat trade," 8

Caccone, Adalgisa, 133–134

California, 77, 79–82, 94, 162, 174–175

California condors, 136–137

Calment, Jeanne, 37

Caloric needs, 17, 31–32, 60

Cambodia, 1, 3, 94

Capacity building, 167

Captive breeding, 86–87, 130–131, 136–137, 146–147, 149–151, 155–160, 164–166, 172–174, 178–179

Captorhinus, 18

Carapace. *See* Shell

Carbon trading, 182–183

Caribbean islands, 20–21, 137

Carr, Archie, 4–5, 169

Carroll, Lewis, 181

Chaco tortoises, 23, 109

Chatuchak Market, 139–141

Chelonoidis, 23, 124, 133, 169. *See also* Chaco tortoises; Red-footed tortoises; Yellow-footed tortoises

Chimpanzees, 34, 37–38

China, 1–3, 12, 65, 93–96, 101–107, 142, 144, 148–149, 170–171

Climate change, 46, 65

Cold-bloodedness, 13, 16–17

Colossochelys, 21

Community-based conservation, 7, 168–169, 176–177

Concordia Turtle Farm, 104–105

Congdon, Justin, 49–52

Congo, 185

Conservation, 3–4, 8–9, 28, 30, 184–186; and tortoise longevity, 53–54, 56–57; and habitat loss, 63, 65, 67–78, 80, 82, 85, 87–88; and food markets, 101, 105, 107; for island species, 117–119, 121–123, 136–137; and pet trade, 145, 149; solutions, 163–180

Conservation banking, 73–74

Convention on the International Trade in Endangered Species (CITES), 140–141, 144–146, 154

Cook, James, 36

Cooters, 104

Coyotes, 83

Darwin, Charles, 35–36, 38, 40, 43, 112–113, 116, 128–129

Darwin Research Station, 130–132, 135, 138, 169–170

Deferred breeding, 40–41

Deforestation, 66, 68, 77, 87–89, 168, 182

Density, 69–70, 159

Desert tortoises, 16–17, 20–22, 42, 51, 60, 77, 79–84, 106, 161–162, 174–175, 182–184

Development. *See* Land development

Diamondback terrapins, 96–98

Diapause, 47–48

Dipsochelys, 114. *See also* Aldabra tortoises

Disease, 81–82, 95, 104, 106
Domed morph, 127, 129
Dry land tortoises, 21–23, 51–52
Durrell, Gerald, 165–166
Durrell Wildlife Conservation
   Trust, 165–166

Ecosystems, 6, 8, 30, 32, 68, 87, 95,
   117, 119, 123, 136–137, 168, 177, 183
Ecotourism, 66–67, 71, 100–101,
   137–138, 177–179, 182–186
Ecuador, 68, 130, 138, 179
Eggs, 13, 20, 41–50, 54, 57, 98, 105
Egyptian tortoises, 176–177
Elephants, 38
Elongated tortoises, 24, 26, 102,
   144
Embryos, 47–48
Endangered species, 28, 140–141,
   145–146, 165, 175, 180
Endangered Species Act, 74, 80
Entombed tortoises, 62–63
Eocene period, 19
Española Island (Hood Island),
   131, 134
Eunotosaurus, 18
Europe, 2, 19, 21–23, 28, 67, 81, 142,
   152–154, 159
European Union, 150
Evolution, 13, 17–21, 135–136, 182
Evolutionary success, 5–6, 40–41,
   51, 59
Extinction, 1, 3–4, 6–7, 28, 54, 56,
   65, 75, 87, 93, 114, 122, 137, 140,
   145–146, 165, 171, 176, 182, 186,
   190

Fady, 167–168
Farm-raised tortoises, 93
Farm-raised turtles, 2, 93–97, 103–
   107
Fertility, 41, 48–54
Flat-tailed spider tortoises, 24–25,
   44, 66, 98–99
Floods, 88–89
Floreana Island, 35, 135
Florida, 62, 76, 168–170
Fong, Jonathan, 148
Food markets, 1–3, 8, 23, 53, 58, 65,
   68–70, 85–86, 90–107, 148, 168,
   186
Forest fragmentation, 60, 72–78,
   86, 166
Forest tortoises, 19–20, 23–25, 33,
   44, 55, 91–92, 102–103, 180, 183–
   184
Forests, destruction of, 66, 68, 77,
   87–89, 168, 182
Fossils, 17–21, 118
Fragmentation. See Forest frag-
   mentation

Galápagos tortoises, 17, 20, 23, 25–
   26, 35–36, 48, 59–60, 67–68, 109–
   112, 115, 123–138, 146, 161, 169–
   170, 178–179
Geochelone, 21, 23. See also Burmese star tortoises; Indian star
   tortoises; Radiated tortoises;
   Spurred tortoises; Star tortoises
Geometric tortoises, 85, 147
Gerlach, Justin, 113–114

Giant tortoises, 5, 8, 17, 20–21, 23, 25–26, 31, 35–36, 48, 53, 59, 108–139, 147, 161, 178–180

Global warming, 46, 65

Gopher tortoises, 21–22, 62, 73, 76

Gopherus, 21, 80. *See also* Bolson's tortoises; Desert tortoises; Gopher tortoises; Texas tortoises

Gorillas, 177–178

"Greek" tortoises, 30, 153

Griffiths, Christine, 119, 123

Griffiths, Owen, 115–119, 136

Growth, 51–52, 60–61

Guides, local, 71

Gulars, 14, 43, 58

Guyana, 169

Habitat corridors, 77–78

Habitat fragmentation, 70, 72–78, 86, 166

Habitat loss, 7–8, 57, 62–89, 163, 165

Habitat specialists, 7–8, 72

Hadrianus, 19–20

Hatchlings, 48, 56, 58

Hawaiian Islands, 120

Head starting, 59, 174–176

Heart, 15–16

Herbivores, 21, 30–33

Herrmann's tortoises, 154

Hibernation, 16–17, 42, 153, 171

Hingeback tortoises, 24, 155

Homopus, 25. *See also* Padlopers; Speckled padlopers

Hong Kong, 102

Human population growth, 64–65, 70–71, 87, 175–176

Hunting pressure, 8, 71, 185

Île aux Aigrettes reserve, 121–123

Illegal trade, 1–3, 7, 86, 98, 140–142, 145–152, 154–155, 158, 164

Impressed tortoises, 24, 44, 70, 157

Incubation, 45–47, 59

Indotestudo, 153. *See also* Elongated tortoises

India, 23, 141–142, 176

Indian Ocean islands, 3, 5, 17, 20–21, 25, 53–54, 109–123, 135–138, 179

Indian star tortoises, 23, 141–142, 147

Indonesia, 1–3, 65–66, 88, 101–102, 104

Isabela Island, 133–135

Island tortoises, 25–26, 31–32, 35–36, 108–138, 178–180

Japan, 2, 93, 142

Jersey Zoo, 166

Jones, Carl, 120–122

Juvik, James, 174

Kaeng Krachan National Park, 183–185

Kenya, 67, 150, 152

Keratin, 14

Kiester, Ross, 174

Kinixys, 24. *See also* Bell's hingeback tortoises; Hingeback tortoises

Klemens, Michael, 151
Korea, 105
Kuchling, Gerald, 171

Lake Park, Fla., 62
Land development, 62–64, 68, 70–
   77, 80, 87–88, 161, 165, 186
Laos, 1, 3, 94
Lawson, Nicholas, 128
Leaf turtles, 104
Leguat, François, 110–111
Lemurs, 66, 100
Leopard tortoises, 21, 26, 175–176
Life history, 50–51
Linnaeus, Carolus (Carl von
   Linné), 27
Longevity, 6–8, 12–13, 35–61
Los Angeles, Calif., 142
Louisiana, 104
Lovejoy, Thomas, 75
Lungs, 15

Madagascar, 2, 22–25, 36, 44, 47, 57,
   66, 69, 86–87, 98–101, 115, 118,
   137, 141, 145–148, 163–168
Malacochersus, 25. See also Pan-
   cake tortoises
Malaysia, 1–2, 102, 104
Male mating aggression, 42–43,
   58–59
Mammals, 37–38, 46–47
Manouria, 19, 20, 44, 102, 157, 183.
   See also Asian forest tortoises;
   Impressed tortoises
Map turtles, 104
Marginated tortoises, 154

Mascarene Islands, 108, 110–112,
   115–123, 179
Mass extinction, 1, 3–4, 6–7, 165
Mating, 8, 13–14, 40–44, 55, 128,
   159
Maturation, 40–41, 50–51, 58
Mauritian Wildlife Foundation,
   121
Mauritius, 110, 112–113, 115–123
McShane, Thomas, 71
Medicines, traditional, 2, 94, 104
Mediterranean tortoises, 22, 28–
   30, 67, 152–155
Meiolania, 20
Melville, Herman, 125–126
Metabolism, 16–17, 31–32
Mortality, 48, 52–53, 56–57
Myanmar (Burma), 1, 3, 23, 66, 68,
   88, 101, 171
Mycoplasma bacteria, 81–82

Nagy, Kenneth, 174
Nepal, 88–89
Nests, 44
Neural bones, 14
North America, 2, 10–12, 16, 21–22,
   49–52, 60, 62–63, 68, 73, 78–84,
   87, 94–97, 103, 106, 142, 159–162,
   174–175, 182
Nutritional needs, 17, 31–32, 51–52,
   60–61, 83–84

O'Brien, Susan, 57–59
Odontochelys, 18
Off-road vehicles, 83, 161
Omnivores, 33

Pacific Ocean islands, 5, 17, 25–26, 109–112, 123–138, 178–179

Padlopers, 5, 25, 31, 157

Painted turtles, 104

Panama, 75–76

Pancake tortoises, 13, 25, 44, 67, 140, 149–152

Paraguay, 109

Parasites, 156, 160

Parham, James, 148

Pelodiscus, 93. *See* Softshell turtles

Pelvis, 5, 13

Pet tortoises, released, 81–82

Pet trade, 2–3, 23, 30, 58, 65, 68–69, 85–86, 94, 98, 102, 104, 139–162, 164, 168, 171, 176–177, 182, 186

Pingré, Alexandre, 111

Pinta Island, 132–135, 169–170

Plastron, 11, 14–15, 43

Pleural bones, 14

Ploughshare tortoises, 22–23, 57–58, 85–87, 141, 146, 163–167, 173

Poaching, 2–3, 8, 30, 45, 58, 67, 69–71, 86–87, 95, 100, 141, 145, 147, 164, 185

Pond turtles, 95

Populations, 28, 53–57, 59–60, 69–70, 73, 98, 159, 166, 182

Potassium, 83

Poverty, 65–66, 68, 70, 87, 185

Predators, 33–34, 45, 48, 55–56, 59, 83

Pritchard, Peter, 168–171

Proganochelys, 19

Protected areas, 58, 64–69, 73, 76, 80, 82, 84–85, 87–88, 121–123, 146, 164, 182–186

Psammobates, 85. *See also* Geometric tortoises

Pyxis, 23–24, 98, 147–148. *See also* Flat-tailed spider tortoises; Spider tortoises

Radiated tortoises, 2, 22–23, 36, 47, 69, 100–101, 118, 140–141, 144–147, 163, 167, 180

Ravens, 83

Red-eared sliders, 93, 95, 104, 160

Red-footed tortoises, 23–24, 68, 140

Reeves' turtles, 104

Reintroduced tortoises, 59, 79–82, 160, 174–176

Repopulation, 59–60, 74

Reproduction, 8, 13, 38–59, 158–159

Reptiles, 13, 16, 44–46, 143, 156

Respiration, 15

Rib cage, 5, 14–15, 18–19

River turtles, 103, 171, 176

Road kill, 63, 78–79, 83, 162

Rock tortoises, 25

Rodrigues Island, 110–112, 117–118, 120, 179

Roofed river turtles, 171

Round Island, 118–120

"Russian" tortoises, 140, 154

Saddle-backed morph, 127–130

San Cristóbal Island, 35, 130

Santa Cruz Island, 123–125, 127, 130, 135, 178

Santiago Island, 35

Sauropsids, 18

Scutes, 14–15

Sea turtles, 3–4, 57, 169

Seed dispersal, 32–33

Senescence, 39, 48–51

Sex, determination of, 45–46, 128–129

Seychelles, 37, 110–111, 113, 179

Shell (carapace), 5–7, 11, 13–15, 18–20, 32, 34, 43, 127–128, 133, 149

Shepherd, Chris, 102

Shi Haitao, 104, 148, 171

Shoulders, 5, 13

Sierra Club, 121

Singapore, 102, 141–142

Singh, Shailendra, 176

Size, 58–59, 129–130

Skeleton, 13–14

Smuggling, 1–3, 7, 86, 98, 141–142, 145, 147, 158, 164

Snapping turtles, 19, 104

Snooke, Rebecca, 152

Softshell turtles, 93, 103, 106, 170–171

South Africa, 25–26, 31, 67, 69, 85, 157

South America, 23–24, 68, 76, 109, 126, 169

Species, definition of, 26–30

Speckled padlopers, 5, 31

Sperm storage, 42

Spider tortoises, 23–25, 44, 47, 66, 98–99, 147–148

Spurred tortoises, 21, 140, 160–161

Star tortoises, 23, 140–142, 146–147, 163

Stewardship, 168

Stigmochelys, 21. *See also* Leopard tortoises

Subspecies, definition of, 26–30

Sustainable development, 70–72

Suurpootjie. *See* Geometric tortoises

Taiwan, 104

Tanzania, 25–26, 67, 150, 152

Temperature-dependent sex determination, 45–46

Testudo, 22, 28–30, 67, 152, 154. *See also* Herrmann's tortoises; Marginated tortoises; Mediterranean tortoises; "Russian" tortoises

Texas, 95

Texas tortoises, 21–22

Thailand, 55, 65, 94, 104, 139–141, 183–185

Thirakupt, Kumthorn, 183

Tonga, 36

Tortoises: classification of, 1–2, 21–26; traits of, 4–5; endangered, 4, 6–8, 140–141, 145–146, 165, 175, 180; history of, 5, 17–21; species of, 5, 21–26, 28–30, 191–194; physiology of, 12–17; niche for, 30–34; longevity of, 35–61; and

Tortoises (*continued*)
     habitat loss, 62–89; as food, 90–107; island species, 108–138; as pets, 139–162; conservation solutions for, 163–180
Tourism. *See* Ecotourism
Translocated tortoises, 63, 82, 179–180
Tunisia, 153
Turtles: classification of, 1–2, 26; endangered, 6; physiology of, 12–17; history of, 17–19; longevity of, 39–42, 48–52, 54–55, 57; as food, 92–99, 101–107; as pets, 144–146, 149, 159–160; species of, 148–149; conservation solutions for, 168–171, 173, 176

Van Dijk, Peter Paul, 171
Vertebrae, 14–15
Vietnam, 1, 3, 65, 94, 104, 170

Wal-Mart, 62
Wanchai, Pratyaporn, 183
Weed species, 7, 72
Whales, 38, 48
White, Gilbert, 152–153; *The Natural History of Selborne*, 22, 153
Wickham, John Clements, 36
Wildfires, 82–83

Yellow-footed tortoises, 23, 68, 76

Zimbabwe, 69
Zoos, 6, 146–147, 165–166, 170–171, 173–174, 182